The Trail of Gold and Silver

T0308916

Timberline Books

STEPHEN J. LEONARD AND THOMAS J. NOEL, EDITORS

The Beast, Benjamin Barr Lindsey with Harvey J. O'Higgins

Colorado's Japanese Americans, Bill Hosokawa

Denver: An Archaeological History, Sarah M. Nelson, K. Lynn Berry, Richard F. Carrillo, Bonnie L. Clark, Lori E. Rhodes, and Dean Saitta

Dr. Charles David Spivak: A Jewish Immigrant and the American Tuberculosis Movement, Jeanne E. Abrams

Ores to Metals: The Rocky Mountain Smelting Industry, James E. Fell, Jr.

A Tenderfoot in Colorado, R. B. Townshend

The Trail of Gold and Silver: Mining in Colorado, 1859–2009, Duane A. Smith

The Trail of Gold and Silver

Mining in Colorado, 1859–2009

Duane A. Smith

UNIVERSITY PRESS OF COLORADO

For
Karen and Mark Vendl,
wonderful friends, fellow Cub fans, and mining historians.

© 2009 by the University Press of Colorado

Published by the University Press of Colorado
5589 Arapahoe Avenue, Suite 206C
Boulder, Colorado 80303

All rights reserved
First paperback edition 2010

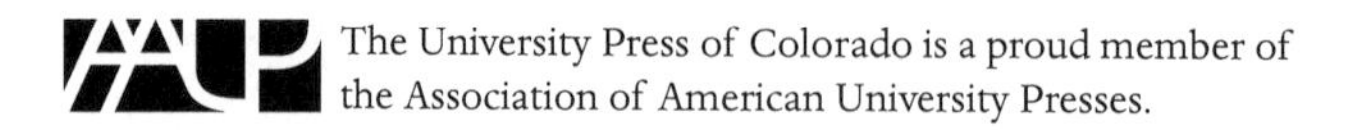

The University Press of Colorado is a proud member of the Association of American University Presses.

The University Press of Colorado is a cooperative publishing enterprise supported, in part, by Adams State College, Colorado State University, Fort Lewis College, Mesa State College, Metropolitan State College of Denver, University of Colorado, University of Northern Colorado, and Western State College of Colorado.

Library of Congress Cataloging-in-Publication Data

Smith, Duane A.
The trail of gold and silver : mining in Colorado, 1859–2009 / Duane A. Smith.
p. cm.
Includes bibliographical references and index.
ISBN 978-0-87081-957-5 (hardcover : alk. paper) — ISBN 978-1-60732-075-3 (pbk. : alk. paper) — ISBN 978-1-60732-011-1 (electronic book : alk. paper) 1. Gold mines and mining—Colorado—History. 2. Silver mines and mining—Colorado—History. 3. Colorado—Gold discoveries. 4. Colorado—History, Local. I. Title.
F781.S64 2009
978.8—dc22

2009019215

Design by Daniel Pratt

Contents

Foreword

Colorado's Clio.
The Homer of the Hills.
The Sage of the Silvery San Juan.
The Monarch of Mining Historians.

Duane Allan Smith has been called many things, but no one can deny he is Colorado's most prolific historian, surpassing even the late, great LeRoy Hafen. *The Trail of Gold and Silver* is Smith's *fiftieth* book. The University Press of Colorado's Timberline Books series, which features the best current work on Colorado as well as classic reprints, proudly presents this master historian's survey of 150 years of Colorado gold and silver mining.

An outstanding teacher as well as an author, Duane has been honored by Fort Lewis College, the Colorado Endowment for the Humanities, and the Carnegie Foundation, and he was named Colorado's Teacher of the Year by the Council for the Advancement and Support of Education. Eternally wearing his crew cut and Chicago Cubs belt buckle, he bounces into the classroom, often with a revolver strapped to his hip and carrying an authentic Civil War musket used by his great-great-great-cousin on Sherman's March to the Sea. An indefatigable lecturer, he amazes his students, who claim he never takes a breath and talks so fast they cannot take notes without a tape recorder.

Born April 20, 1937, in San Diego, Duane is the only child of Ila Bark Smith, a schoolteacher, and Stanley Westbrook Smith, a Navy dentist whose life his son recorded, including his adventures as a Japanese prisoner of war during World War II. The family vacationed in Colorado and Duane fell in love with the Highest State. He returned to complete his B.A., M.A., and Ph.D. in history at University of Colorado at Boulder.

His dissertation was his first book, *Rocky Mountain Mining Camps: The Urban Frontier* (Bloomington: Indiana University Press, 1967). This important work focused on the urban nature of the mining West. Drawing on Richard Wade's pathbreaking book, *The Urban Frontier*, Duane reshaped the way scholars approach the American West. Disagreeing with Frederick Jackson Turner's venerable Frontier Thesis, Smith argued that the mining West was an urban frontier shaped by groups of miners and those mining the miners rather than a rural effort of rugged individuals. His second title, coauthored with Carl Ubbelohde and Maxine Benson, *A Colorado History*, has been through many editions and remains a widely used text, and it provides hearty competition for another text of which I am a coauthor with Carl Abbott and Steve Leonard, *Colorado: A History of the Centennial State*. Smith followed *A Colorado History* with one of his best sellers, *Horace Tabor: His Life and the Legend* (Boulder: University Press of Colorado, 1973). Probably his most important mining books are the previously mentioned *Rocky Mountain Mining Camps*, *Rocky Mountain West: Colorado, Wyoming and Montana, 1859–1915* (Albuquerque: University of New Mexico Press, 1992; in the prestigious History of the American Frontier series), and *Mining America: The Industry and the Environment, 1800–1980* (Boulder: University Press of Colorado, 1993), which Duane dedicated to Colorado's U.S. senator Gary Hart, whose La Plata County senatorial and presidential campaign Smith managed. Duane's primary interest is Colorado mining, as he explained to *Contemporary Authors*: "Probably the most important motivation for research on mining camps was the desire to uncover a more realistic and honest history. I believe that history is not dull, but writers and teachers make it so. Therefore, my goal as a writer and teacher is to make history come alive, to hook people on history."

Besides his interest in mining, Smith is also noted for his love of bears, especially the Chicago Cubs, and his love of cats. He now "owns" six but in 2007 hit a peak of nine felines. He usually works these creatures into his

Colorado, Civil War, mining, and baseball history courses at Fort Lewis College in Durango. Duane has been a professor of history and Southwest Studies there since 1964, when he received his Ph.D. from the University of Colorado at Boulder, where he worked with professors Robert Athearn and Carl Ubbelohde.

Duane has also served as Preacher Smith at Durango's First United Methodist Church, where he teaches a Sunday School class in Bible baseball, a game he invented. He sings in the choir, along with his wife, Gay Woodruff Smith. They met in a history class at CU, where Gay was the top woman graduate and Duane the 180th overall in the class of 1959. Gay helped him along then, editing his work, and she has done so ever since.

Of her husband, Gay says, "Inside this well-known and knowledgeable Western historian beats the heart of an irrepressible and mischievous little boy." Their daughter, Laralee, describes her father as "fun to be around," even though he raised her by reading history to her, hooking her on the *Little House of the Prairie* series, and dragging her to historic sites—from Civil War battlefields to mining ghost towns.

An active citizen, Duane served on the La Plata County Historical Commission, which he chaired, and on Colorado's National Register Review Board. He remains chair of the Durango Parks and Recreation Board and proudly leads walking tours along its Animas River Trail. He also served on and chaired the Durango Historic Preservation Committee.

Professionally, Smith was a 1960 founding member of the Western History Association and has attended every meeting. He is one of the six 1988 founders and the third president of the Mining History Association, which has grown into an international organization with 300 members worldwide. Smith is sheriff of the Durango Posse of Westerners, which he founded in 1976. He also co-founded the La Plata County Historical Society in 1974, and he remains an active board member. Duane was instrumental in restoring the Animas City Schoolhouse, where he delights in conducting tours for fourth- and fifth-grade Colorado history students.

Among many distinctions, Smith has been a fellow at the Huntington Library in San Marino, California; an advisory board member and historian of the Durango & Silverton Narrow Gauge Railroad; a Colorado centennial-bicentennial commissioner; and a member of the La Plata County Democratic Executive Committee. He worked on La Plata County political campaigns for governors Roy Romer and Richard D. Lamm, with whom he coauthored *Pioneers and Politicians: Profiles of Colorado's Governors* (reprinted in 2008 by Fulcrum in Golden). For his achievements as an author, teacher, public speaker, and civic activist, the Denver Posse of Westerners awarded him its prestigious $1,000 Rosenstock Prize for Outstanding Contributions to Western History.

Smith has written the history of Fort Lewis College and served three times as its graduation commencement speaker. He is a former director of the college's Center for Southwest Studies, which in 2000 opened a fabulous new circular stone complex of offices, classrooms, a library, and exhibit space. A jogger and athlete, Smith still plays softball. He coached the Fort Lewis girl's basketball team in 1968–1969—the only Fort Lewis College team to be a winner that season. For this and other service to the school's athletic program, he was inducted into the Fort Lewis College Athletic Hall of Fame in 2004.

Smith contends that mountains and mining distinguish our region. Leaving earlier Native American and Hispanic activity to other scholars, Smith focuses on English-speaking settlement. He begins with the great mining rushes and is fascinated by "the magic of the mountains." He displays a mining man's reservation about Colorado's flat eastern parts and even approaches the Queen City of the Mountains and Plains with skepticism. He laments with Mark Twain the "monotonous execrableness" of the plains whose only salvation is their "mountain vistas."

Duane Smith's passion for primary sources and mastery of secondary sources shines once again in *The Trail of Gold and Silver*, a superb overview of hard-rock mining that ponders past failures as well as successes. From the bonanza at Cripple Creek to Summitville's environmental disaster, Colorado mining has had ups and downs as high as Mount Elbert and as deep as the Royal Gorge.

As the pacesetter for other Colorado historians, Duane is forever urging us all onward and upward. Yet, I wish he would stop his annoying practice of waking us up at 6:30 A.M. with questions and suggesting we get up and start writing. I once made the mistake of coauthoring a book, *Colorado: The Highest State*, with this man. He had the nerve to send me his half a month later—and then asked where mine was. Such flaws notwithstanding, let us wish Duane, as he wishes all on his phone message in his own cheery, wide-awake voice, "Top of the Day." As you will find in the following pages, Duane Smith is still profitably mining Colorado's hard-rock past.

—THOMAS J. NOEL, 2009

The Trail of Gold and Silver

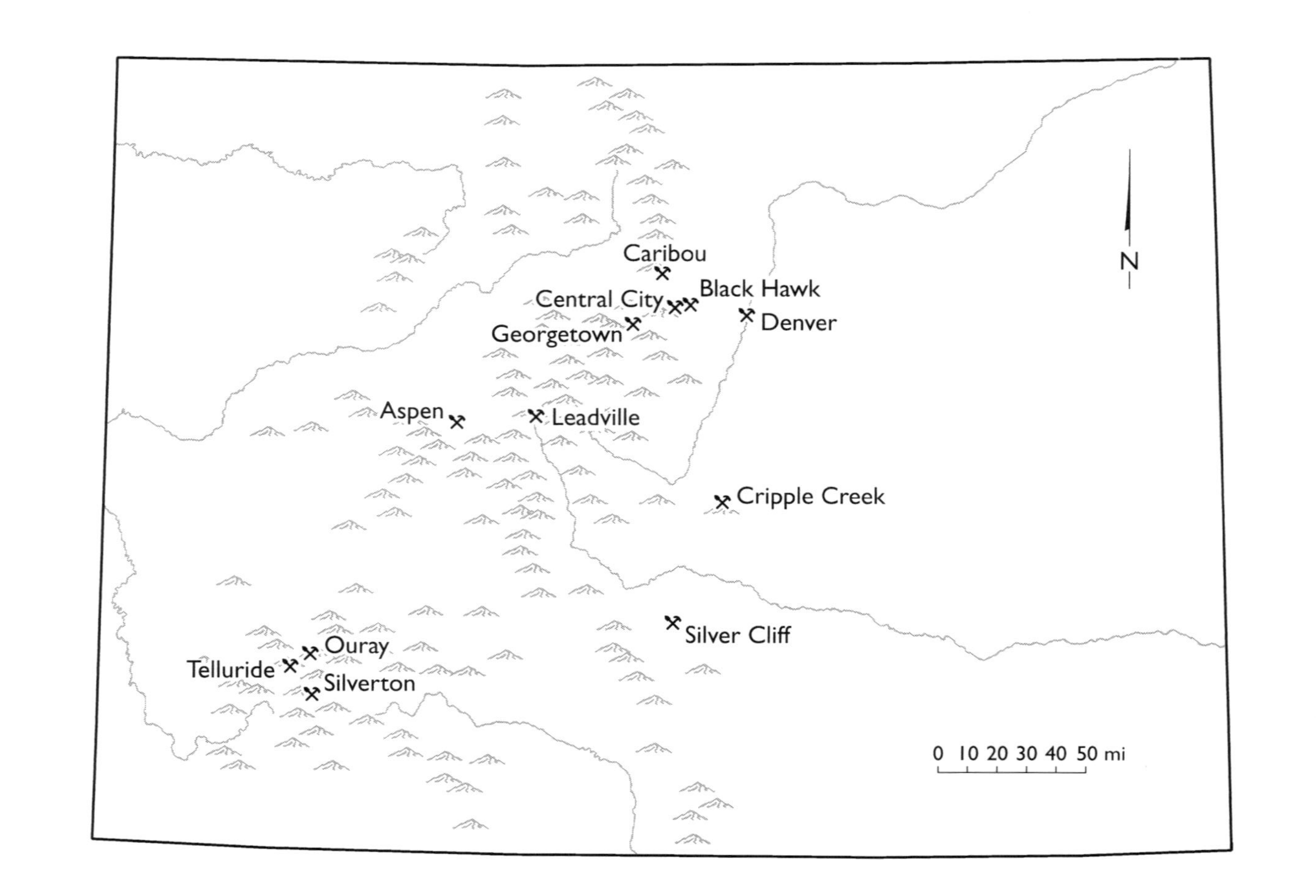
N
Caribou
Central City
Black Hawk
Denver
Georgetown
Aspen
Leadville
Cripple Creek
Silver Cliff
Ouray
Telluride
Silverton
0 10 20 30 40 50 mi

GOLD! GOLD!! GOLD!!! GOLD!!!!

—*KANSAS WEEKLY PRESS*, SEPTEMBER 4, 1858

Prologue

"Gold! Gold in the Pike's Peak Country," shouted newspapers throughout the Midwest in the late summer and fall of 1858. The news spread over a country still stirred by the tremendous excitement of the California gold rush a decade earlier. Wonder of wonders, had it happened again? Trapped in the worst depression in living memory, many Americans hoped against hope that it had. It would take nearly a year to sort the rumors from the reality.

The golden saga, though, did not begin here. Rumors of mineral wealth in the area dated back almost to Christopher Columbus's arrival in the New World in 1492. Teased by finding a little gold, the Spaniards in Central and South America hit the bonanza in the early sixteenth century, when they conquered the astoundingly wealthy Inca and Aztec empires. Lured by stories of even more fabulous wealth in the "Seven Cities of Cibola," with their streets of gold and tinkling silver bells, Francisco Vazquez de Coronado led an expedition northward up the Rio Grande to the "Tierra Nueva." This wandering, desperate search of 1540–1542 turned up neither gold nor silver, and the discouraged Spaniards eventually trudged back to Mexico.

Time eventually obscured the reality of what Coronado had seen and endured; a generation slipped away before old stories of mythical gold brought the ever-eager Spaniards back to the upper Rio Grande country. By the early 1580s, they were venturing northward, searching for "mines." Despite reports of "extremely rich veins, all containing silver deposits," no rich mines were found and no rush developed.

Always hopeful, despite a notable lack of success, they did not give up. Next came Juan de Oñate and, with him, permanent settlement in 1598. Except for the twelve years following the Pueblo revolt of 1680, a permanent European colony—Nueva Mexico—persisted on the lonely northern fringe of Spain's New World empire. However, the colony struggled to survive. No new golden or silver kingdoms were found, and little evidence of precious metals was uncovered locally. Nevertheless, the dreams and stories of wealth never flickered out completely.

Sometime in the mid-seventeenth century, someone ventured across the miles—from Albuquerque or Taos or Santa Fe or places in between—toward the mountains to the north. There they apparently found deposits of placer gold in the streams, and perhaps outcroppings of minerals on the mountainsides. They also found that the indigenous inhabitants, the Utes, were not terribly happy about this incursion into their domain; the Utes posed a threat to every intruder. No frenzied media reports of gold from the isolated, scattered settlements along the Rio Grande tempted locals northward to make their fortunes. Only a few New Mexicans and Spanish officials knew about those rumors.

In contrast, the French, in the Midwest and the St. Lawrence Valley, had been hearing rumors of "rich mines" in the Rocky Mountains for years. In 1702, a party left Illinois to see mines the Indians had told them about; this venture was followed in 1723 by a report of copper and silver mines. Sometime after 1739, a group journeyed westward but apparently failed to find any treasure. A 1758 map vaguely located a purported gold mine on the Arkansas River.

Not to be outdone, the rival English claimed that the country to the west was "full of mines." Some of these ambiguous, exaggerated reports probably rested on precariously little fact, but the ever-present hopes and rumors spawned legends of lost mines, buried gold, and a lone survivor of an Indian attack who had carried a cryptic map back to the settlements and then promptly died before furnishing any further information. These treasure tales echoed down the decades, teasing and tempting each new generation.

The Spanish settlers, who lived closer to the locus of the legends, had by no means abandoned their quest. Juan Maria Antonio Rivera led two 1765 expeditions into the rugged, high La Plata and San Juan mountains, searching particularly on the first one to meet a Ute to guide him to a reputed silver deposit. A Ute, Wolfhide, had appeared earlier in Santa Fe with wire silver ore, and the governor was determined to find the source. So out went Rivera—and found his prey elusive, although he skirted and perhaps probed both the San Juan and La Plata mountains. He eventually reached the Gunnison River before returning to Santa Fe. Following Rivera's expedition, New Mexicans apparently prospected and even spent time mining in the La Plata Mountains and farther into the high San Juans. Not wishing to pay the "Royal Fifth" of all they found to a far-away king in Spain, they left no trail of records for local officials—or historians—to follow.

While Americans struggled for their independence along the Atlantic coast, the governor of New Mexico read a report from the Dominguez/ Escalante expedition:

> The Rio de la Plata flows through a canyon in which they say there are veins and outcroppings of metal. . . . The opinion formed previously by some persons from the accounts of various Indians and of some citizen of this kingdom that they were silver mines, caused the mountain to be called Sierra de la Plata.

On a hot August 9, 1776, Velez de Escalante jotted rumors about mining in his journal, as his party rode past the south end of those "silver" mountains. The missionaries in this group were not looking for gold or silver; rather, they were seeking "bearded Indians," converts for their Catholic faith, and an overland route between Santa Fe, New Mexico, and Monterey, capital of Spanish California. Still, any hints about precious metals were considered worth recording.

Change came slowly to New Mexico, but come it did. Far to the east, the newly victorious "Americans" began to move westward, searching for, among other things, the gold and silver that had helped motivate the establishment of the first English settlement at Jamestown, Virginia, back in 1607. Neither determination nor luck had brought success, but the hope of instant wealth would not die.

By the early 1800s, these foreigners had reached New Mexico, much to the dismay of Spanish officials. Particularly after hearing of the sale of the Louisiana Territory to the United States in 1803, the Spanish worried what the future might hold: That purchase placed the aggressive outsiders right next door. New Mexico could expect little military assistance, or any other help, from its mother country, which had become merely a pawn in the English-French struggle to dominate Europe.

The New Mexicans had every right to be concerned, because Americans, drinking deep from the cup of manifest destiny, had visions of controlling North America—and everything else they could lay their hands on. Was it not, after all, their God-given right to bring democracy and their advanced political and social institutions to more backward peoples? They could see little reason to doubt the righteousness of their endeavors.

Before long, the threat became reality. Wanting to know what he had purchased, President Thomas Jefferson sent out two expeditions, one led by Lewis and Clark that went up the Missouri River, and one led by Zebulon Pike that went across to the Rocky Mountains. Pike's 1806 mission was to go into the Southwest and find the source of the elusive Red River, the southern purchase boundary.

Pike reached the Rocky Mountains, tried to climb the peak now named after him, and eventually journeyed across the Sangre de Cristo Mountains into the freezing, winter-locked San Luis Valley. There he finally built a stockade on what he thought was the Red River. All his party's movements were known to the Spaniards in New Mexico, who had earlier tried but failed to intercept the expedition on the Great Plains. They logically waited until spring to have a patrol take the Americans into custody and escort them to Santa Fe. Pike's claim that the Rio Grande was in fact the Red River did not sway New Mexican officials in the least, and the group was taken, under guard, to Chihuahua, before finally being dumped back into American territory in June 1807.

During his sojourns, Pike found no gold, but he met someone who had: James Purcell, a trapper who had also wandered into Spanish territory and found himself forced to remain there. The former Kentuckian "assured me that he had found gold on the head of the La Platte [sic], and had carried some of the virgin mineral in his shot-pouch for months." Purcell steadfastly refused to tell Pike exactly where he had found the gold, because he believed the site to be in American territory.

This tidbit was included in Pike's report of 1810, but it did not stir much interest. Troubles with England were about to erupt into another war, and New Mexico was still a long way from the advancing western frontier.

Even less well known was the report of St. Louis resident Regis Loisel, who had been sent out by the French governor, back in April 1803, to examine the western territory. Upon his return, he found that the United States now owned the region; Loisel nonetheless filed an 1804 report that, among other things, claimed he had found gold.

Though the scattered, fragmentary evidence was mounting, no one had yet put all the pieces together. Nor had there been a gold strike in the last century in what was now the United States to whet the appetite of the adventuresome. However, the pace of westward expansion was quickening. With the war over, Americans moved beyond the Mississippi Valley. The profits of the fur trade beckoned them into every Rocky Mountain nook and cranny. That the Spanish government (or anyone else) continued to be hostile did not bother the aggressive Americans much.

Meanwhile, the Mexican people threw off Spain's yoke, and declared their independence. Their government then made a move it would later regret bitterly: it opened Texas, California, and New Mexico to Americans, hoping to build up settlement, increase trade, and provide better defense against marauding Plains Indians and other tribes. Over the Santa Fe Trail the invitees came, ready to trade, and moved into the mountains looking for beaver. Before long, gold rumors began to filter back East from the Rocky Mountains, though few of the tale-tellers brought any actual gold or silver with them. Trapper James Cockrell thought he found a silver mine in 1823, for instance. Four years later, an expedition found "ore" and laboriously packed it back to St. Louis by horse—only to have an assay crush their hopes.

Reality seldom puts rumors to rest, though, and America's first real gold rush, in western Georgia, gave them all new life. In 1828–1830, and the following years, the eager and excited hastened to the mining communities of Auraria and Dahlonega, to find their fortunes "without working." They camped and prospected, they panned, they dug, they burrowed, they sweated in the heat and humidity; they even watched slaves work claims for their masters. Most came away disappointed, having discovered only astonishingly hard work and meager returns. As one old miner moaned, "I've never worked so hard in my life to get rich without working." Georgia's was less than a huge stampede by later standards, and only encompassed a fairly small area, but it was the first American gold rush . . . and it revived the dreams and hopes that other golden troves awaited the fortunate adventurer.

During the decade of the 1830s, both Indians and Mexicans brought gold to Fort Vasquez and Bent's Fort, located beyond the foothills on the South

Platte and Arkansas Rivers. William Bent claimed that the Indians had long known of the presence of gold in the country. It was, however, in the natives' best interests to keep quiet; a gold rush could potentially destroy their way of life forever.

A party of trappers and traders headed out of New Mexico to the Vasquez Fork of the South Platte on a prospecting tour, first in 1833, then during the winter of 1834–1835. According to them, they made a "paying strike," but did little about it. Another trapper, lost in the Rockies in 1835, found some gold specimens, finally regained his bearings, and returned to New Mexico. He excitedly organized a prospecting party, but never could relocate the strike site.

By the late 1830s, trappers along the Missouri River were showing gold "flakes" to curious locals. Possibly truth-based claims were often accompanied by very tall tales: One that persisted for many years held that the Arapahos used "gold" bullets in a desperate fight with a rival tribe. In the 1840s, along with increasingly frequent oral reports, the reading public began to encounter comments such as one in Josiah Gregg's classic *Commerce of the Prairies*. He observed that there existed an "extensive gold region about the sources of the South Platte; yet, although recent search has been made, it has not been discovered." Gregg included a chapter on New Mexico mining, which unambiguously indicated mineralization in the region. Likewise, Rufus Sage mentioned gold several times in his *Scenes in the Rocky Mountains*. He also reported that "Mexicans mined gold in the Sangre de Cristo Mountains," and repeated other stories about trappers or hunters finding gold. His comment that "doubtless very rich mines" existed may have piqued some interest, but few were stirred to hasten west seeking gold.

William Gilpin, later a famous figure in Colorado's saga, went west on several expeditions in the 1840s. He, or one of his party, found gold. Excited about what he had seen and what he believed the future held, Gilpin gave speech after speech in the 1850s about precious metals locked in the central Rocky Mountains. For him, no question remained: "The facts then and since collected by me, are so numerous and so positive, that I entertain an absolute conviction, derived from them, that gold in mass and in position and infinite in quantity, will, within the coming years reveal itself to the energy of our pioneers."

Dreams, rumors, hopes, untraceable gold deposits, and tall tales—they all became reality on the brisk morning of January 24, 1848, in California, when James Marshall found yellow "specks" in the millrace he was building for entrepreneur John Sutter on the South Fork of the American River. That moment changed the destiny of the United States and shaped the future of the American West.

"Boys, I believe I have found a gold mine," Marshall told his fellow workers, in one of the most understated pronouncements ever heard. By mid-March, despite Sutter's attempts to keep the discovery under wraps, the news had leaked out; nothing travels faster than the announcement of a gold strike. "Gold has been discovered in the northern Sacramento District about forty miles above Sutter's Fort," San Francisco's *California Star* (March 18) calmly reported. Readers were far less placid: some immediately hurried to the discovery site. By mid-May, the *Star* editor intimated that "El Dorado" had been discovered: "Parties who have penetrated and traversed the region in which originally gold was found give encouraging reports of its increase in quantity—undiminished in purity" (May 27). As another paper more emotionally put it, California "resounds with the sordid cry of '*gold, gold, GOLD!*' " Within months, the forty-eighters were scattered throughout the Sierra Nevadas, over what became known as the motherlode country, looking for free (placer) gold in the streams and rivers. They found abundant gold almost everywhere they prospected. Jerry-built mining camps sprouted up almost overnight, and near the diggings, eager opportunists arrived to "mine the miners."

The news quickly traveled to all points of the globe, and a worldwide rush ensued. The 100,000 or so forty-niners who arrived the next year by ship, wagon, and horse took up digging, panning, and sluicing, not only enriching themselves but also nearly upsetting the world's economic system. Never before had so much gold been pumped into it in such a short time. Although the forty-niners had to work much harder than their predecessors of only a year ago to get their "poke," the 1849 rush dashed on into the 1850s. Among those journeying west in 1850 was a group of Cherokees from Indian Territory, who cut across the eastern foothills of the Rockies to intercept the well-traveled Oregon Trail. They did a bit of panning along the way, but found nothing interesting enough to make them pause on their way to El Dorado.

During the California rush, William Gilpin kept telling listeners about the great future for potential wealth in the Rocky Mountain region. Anyone interested could also read numerous accounts of gold being found there. Several forty-niners told of panning for gold on their way west, including one who claimed to have purchased $2,000 worth of gold dust from an Arapaho Indian who had dug it around Ash Hollow, Nebraska!

The city of St. Louis, along with its newspapers, fancied itself the gateway to and speaker for the great West beyond, and the news editors were not shy about publishing stories of gold. The *Evening News* (July 9, 1853), for instance, gushed that the South Platte and South Park areas "possess[ed] extensive gold fields." Other papers heralded golden news as well. A letter from California, printed in the Little Rock *True Democrat* in December 1857, spoke of "gold prospects on the upper waters of the Arkansas and Platte" that were "better" than any in California.

Not that the *Evening* News was taken in by every story. It reported on May 14, 1855, that the "existence of rich gold diggings on the head water of the Arkansas river, still continues, though slightly modified by adverse" reports. Its rival, the *St. Louis Intelligencer,* on July 16, told its readers solemnly that another letter writer "is convinced that it is all a hoax."

An interesting aspect of many such reports is that they placed the discoveries in some of the very spots where gold and silver were later actually found. True or not, stories, tales, and rumors of discoveries only heightened the interest. The California strikes kicked off a generation, and then a second generation, of mining frenzies that eventually carried prospectors and miners throughout the Rocky Mountains and southwestern deserts, on into Canada, and north to Alaska. Before the rushes ended, an epic saga had been written across the western North American continent.

All of this excitement did not go unnoticed by the Plains and mountain Indians. Although they found relatively few gold seekers trespassing on their lands, they saw throngs of folks in rambling wagon trains going over the Santa Fe, Mormon, and Oregon Trails to Utah, California, New Mexico, and Oregon. No one could outrun cholera, or the other diseases, that came west as well; such white man's diseases nearly eradicated some of the individual Indian bands. Further, the buffalo, that lifeblood of the native peoples, was decreasing in numbers. While not solely the fault of the traveling pioneers and hunters, the tribes knew who to blame.

Of course, these westward-forging pioneers did not think of themselves as interlopers, but rather as fulfilling their own and their country's manifest destiny. That they disrupted and threatened the existing inhabitants' way of life probably did not enter the minds of most of them. Nor were these travelers and would-be settlers at all happy about the fact that the native peoples stood in their way. All told, an emotionally charged tinderbox waited to explode along the trails, trapping newcomer and older resident alike. A predictable result ensued: a recurrence of the war that had started back in 1607 between two peoples who each desired to call this land home.

The European and American interlopers had climbed the Appalachian Mountains, surged across the Midwest, and now crisscrossed the trails west.

Then came the California gold strikes—never before had anything like it drawn such multitudes westward.

Washington responded haphazardly, first putting troops in the field in the 1820s along the Santa Fe Trail, and then, in the 1830s, building a string of forts to guard it and other trails. The rushes to Oregon and California only heightened the tensions, and in 1857 another effort was made to resolve the ongoing conflict. The army sent troops into the field that summer, and along with them went a Delaware scout, Fall Leaf. Somewhere along the campaign trail, he acquired some gold. The details of just where and how remain hazy, but the important thing was that he carried the gold back to Lawrence, Kansas, where it fired many imaginations in that depressed farming community. The crash of 1857 was in full swing, and an anxious population was more than ready for hints of possible salvation.

As the fall of 1857 settled over the plains and mountains, the Ute, Arapaho, and Cheyenne tribes enjoyed one last season of quiet before the dawn of a new era. Shakespeare warned, "All that glitters is not gold," but 300 years of Rocky Mountain rumors, legends, expeditions, and mining attempts convinced many that there was indeed gold out there in the foothills and mountains, despite the vanishingly small amount of hard proof available. In 1857, three factors came together for the first time that motivated people to turn rumor into reality.

First was the lure. California abundantly proved that gold *was* out there, awaiting the fortunate or determined discoverer. The fact of gold in the Sierra Nevadas made it seem only logical that it would be found everywhere, or, at the least, that there would be a couple of new Californias. Furthermore, some forty-niners had returned home with unslaked gold fever and the skills and experience needed to open and develop a new gold district.

Second, thanks to Uncle Sam, the way west had become better marked and safer than ever before. Through generous government policies and pioneers' determination, settlement had crawled to within thirty days' travel time of the central Rockies.

Third, there was a mighty push from economic conditions. By 1857, the United States had sunk into a deep, desperate depression that hit urbanites and rural folk alike. Americans had never seen anything like it; it was the worst of times in national memory. Owners lost their homes and farms; economic disaster faced many other Americans (particularly Midwesterners and Northerners) as businesses failed, jobs disappeared, and banks collapsed, taking their uninsured depositors' money with them. Farm prices plummeted, railroads declared bankruptcy, and disheartened, discouraged folk sought work of any kind—and there seemed none to be had. Mother Nature had no mercy, either: recurrent, seasonal fevers and agues swept the Midwest. All in all, Americans could point to little in 1857 that cheered them. The American dream of a better tomorrow appeared to be slipping away as the older agrarian America faded before advancing industrialization.

To add to their woes, perspicacious citizens could see an oncoming clash between industrial Northerners and slave-owning Southerners over a variety of issues, including states' rights, sectional economies, political power and political parties, and the ever-emotional slavery question. Agricultural Southerners enjoyed gloating that they were weathering the depression better. In Washington, the halls of Congress echoed with recriminations, threats, and outright hatred. It seemed that the country and its leaders had lost their direction and that the future of America teetered in the balance. Some radicals went so far as to assert that separation offered the only answer.

Americans looked for something, almost anything, to pull them out of this mess. What that might be they did not know, but prayed to find out, as the winter of 1857–1858 settled over the land.

The Gold is there, 'most anywhere.
You can take it out rich, with an iron crowbar,
And where it is thick, with a shovel and pick,
You can pick it out in lumps as big as a brick.

(*Chorus*)
Then ho boys ho, to Cherry Creek we'll go.
There's plenty of gold,
In the West we are told,
In the new Eldorado.

—*ROCKY MOUNTAIN NEWS*, JUNE 18, 1859

1

Pike's Peak or Bust

A gloomy winter slipped past, but gave way to a spring of 1858 that was not much brighter economically. The flowers bloomed, but too briefly to take most people's minds off the trials of daily life. Still, for two groups of people, spring finally brought the chance to follow their cherished dreams of gold beyond the western horizon. Gold, they had every right to believe, was waiting out there. One group, in Indian Territory, prepared to travel the relatively short distance to the foothills of the central Rockies, where some of them had found gold back in the California excitement of 1850. The other group, in Lawrence, Kansas, also knew that gold existed to the west, because the army scout Fall Leaf had brought some back from the military campaign there the year before. Unfortunately, he did not know its source, so they had to decide where to start searching.

As the two parties provisioned and prepared to travel, a betting person would have put money on the success of the Russell party out of Indian Territory. Its members had the most experience: Some had come from the

Georgia gold fields, and others were from California—where, like many of their fellow rushers, they had met with little success. Their lack of success eventually forced them to return home, but gold fever still burned in them.

The initial discovery of placer gold, near where Denver one day would be, dated from June 1850, when a Cherokee party traveling to California panned some color near a small stream later known as Ralston's Creek. As times worsened in the late 1850s, the siren promise of those few flakes of gold intensified—if they could just relocate the place.

Veteran Georgia and California miner William Greenberry Russell, known as "Green" Russell, emerged as the major mover in the early days of Colorado mining. Related by marriage to the Cherokee Indians, he had heard about the 1850 discovery of "color." That intrigued him, as did the idea of seriously prospecting in the Rocky Mountain foothills. Correspondence during the winter of 1857–1858 between Russell and the Cherokees laid the groundwork. Russell, his two brothers, and six others headed west to Indian Territory in February 1858. There they gathered supplies and other needed equipment, and found more men who agreed to join the expedition.

Eventually, the party traveling west included the Russells, some Cherokees, and two groups of Missourians—104 men in all, according to Luke Tierney, one of the early fifty-eighters. After intersecting the well-known Santa Fe Trail, they traveled to Bent's Fort, then up the Arkansas River almost to Fort Pueblo before turning northwestward toward Cherry Creek. By May, the Russell party had reached the site of the future Denver and the area in which gold had been found eight years earlier. By late June, all the groups had ended up together.

Disappointment, not gold, was their reward. Over the next couple of weeks, many discouraged men decided to return home. Finally, only thirteen men—all Georgians, led by Green Russell—resolved to prospect further. They moved northward along the foothills for about thirty miles.

Unbeknownst to the Russell group, another party had had the same idea. Fall Leaf's little sample had stirred up a good bit of interest among Lawrence folk. The long winter evenings gave locals plenty of opportunity to sit around and talk, and dream, and plan. Still, despite persistent attempts at persuasion and promises of rewards, Fall Leaf refused to guide anyone west to the gold regions. His stubbornness was probably wise: he had no idea where the gold had actually been found, and the Lawrence party going west may have seemed too small for safety.

Undeterred by such a minor setback as lack of a guide, when weighed against the expected rewards, about thirty men decided to try their luck at finding the site. In late May, they too set off westward toward Pike's Peak, the only well-known and prominent point in the region. Eventually, joined by a few others, their party grew to a total of forty-nine, including two women and a child.

As they headed for Pike's Peak, they met some members of the discouraged Russell party, but the Lawrence group determinedly continued. Camping near Garden of the Gods, they enjoyed the magnificent scenery, and several small parties climbed Pike's Peak (including the first woman, Julia Holmes, to do so). They found no gold, however, and moved north to Cherry Creek.

At Cherry Creek they joined some members of the Russell party who had not yet given up. They prospected but found no gold; hearing of gold diggings down near Fort Garland in the San Luis Valley, the group turned southward. They found old diggings, but yet again, prospecting failed to yield any gold. By that time, some discouraged members of the party had given up and returned home or gone on to Taos, New Mexico, to check out that area (Holmes among the latter). Then, at last, the group that had remained near Fort Garland heard exciting news.

The Russell party had returned to Cherry Creek and finally found some placer diggings. One of the party, William McKibben, who had a fine sense of history, described what happened when Green Russell returned to camp and "gave us the astounding intelligence that he had discovered a mine where we could realize $15 per day":

> Our joy knew no bounds, we huzzaed, whooped and yelled at the prospect of being loaded with gold in a few months, and gave vent to any amount of hisses and groans for our apostate companions that were making all speed for home. We congratulated ourselves, sir, that we inaugurated a new era in the history of our beloved country.[1]

At this point, the "Paul Revere" of the story arrived at the camp. While at Fort Laramie, trader John Cantrell had heard about men prospecting on the South Platte and decided to visit them on his way home to Westport, Missouri. Arriving on July 31, he stayed for a few days, secured some gold dust, and started home.

By late July, even before Cantrell reached home, rumors of gold discoveries reached newspaper offices along the Missouri, traveling via that mysterious mountain "telegraph." Both Leavenworth, Kansas, and Kansas City, Missouri, papers printed highly exaggerated stories based on reports "direct from the mountains." The former claimed that 500 miners were making, on the average, $12 per day.[2] The latter cut that figure to 150 miners making $8 to $10 per day. Considering the times, both stories must have caught readers' attention. The *Journal of Commerce* could not restrain itself: "The gold discoveries are creating great excitement in the mountains."[3]

Meanwhile, by the time Cantrell reached the Missouri River towns, his imagination had long since taken wing. Initially he reported that the Russell party took out $1,000 in ten days; even more encouragingly, he "thinks if properly worked," one man could make $20 to $25 per day. That was a month's wage for many people, who earned a dollar or less per day, and farmers did not make more than a few hundred dollars a year. No wonder people thought El Dorado had been found! For the depression-locked Missouri River towns and farmers, no news could have been better.

Every day the story improved. Cantrell claimed that he had "traversed about 70 miles of country and every stream," and on every one he prospected he had "found gold." Thanks to the telegraph, the news spread relentlessly throughout the country, although never as fast as the exaggerations and speculation, which exploded like a prairie wildfire.

> Gold discovered at Pike's Peak . . . Parties are starting from various points for the new diggings.
>
> *Kansas Chief* (White Cloud, Kansas), September 9, 1858

> Gold at Cherry Creek . . . Gold found in all places. A new gold fever may be predicted as plainly at hand. Eastern slope of the Rocky Mountains richly treasured with Gold.
>
> *St. Louis Democrat,* quoted in
> *The New York Times*, September 20, 1858

> The Star Salon has a large specimen of gold bearing quartz directly from the Pike's Peak mines. It seems to be very rich, and the sight of it is not calculated to allay "gold fever."
>
> *Leavenworth Journal*, September 14, 1858

> Those best acquainted with mining will no longer doubt . . . gold abounds in the Streams.
>
> *The New York Times,* September 28, 1858

Perhaps the most enthusiastic was the Elwood, Kansas, *Kansas Weekly Press* (September 4, 1858):

> Gold! Gold!! Gold!!! Gold!!!!
> Hard to get and Heavy to Hold
> Come to Kansas!!

Meanwhile, some members of the Russell party ventured into the mountains and came back empty-handed. The entire group then took an extensive prospecting tour north along the Front Range well into Wyoming, without success. When the members of the party returned to Cherry Creek in late September, they were stunned to find members of the Lawrence party there, as well as others, including some traders. Still others soon joined them, the first of the fifty-eighters who had hurried west because of Cantrell and the newspapers' overblown accounts.

There was more than one way to make money in a gold rush, and the Lawrence party set about doing so by laying out Montana City, a site on which they built a few cabins. Not everyone thought this the best site, though, so, after organizing a town company, others laid out another "city"—St. Charles—and set out for home, leaving one member to protect their paper city. Thus did urbanization arrive at the future site of Denver.

At that point the remainder of the Russell party decided to split up. Some went home to get supplies for next year's prospecting; others went to Fort Garland to obtain winter supplies, and then rejoined the few who had stayed at Cherry Creek. The Cherry Creek numbers were increased by early-arriving rushers and the former residents of Montana City. They eventually called their site Auraria, after the hometown of the Russell brothers, who were telling folks in their hometown, according to the *Dahlonaga Signal,* that "prospects for gold in that country are quite favorable."[4] The Aurarians founded the Auraria City Town Company in late October 1858—a month before the Denver City Town Company formed on November 22, 1858, thus becoming the first permanent community within what is now metro Denver.

All this excitement resulted from perhaps $500 worth of gold having been found over the summer, most of it in the Cherry Creek region. That averaged out to less than $5 per person in the Russell party and even less if one figures in the Lawrence group, which had not found any gold. Their consolation prize was a wonderful tourist excursion through some of Colorado's beautiful country!

Not surprisingly, urbanization closely followed the gold seekers. Miners theoretically would find enough gold to pay for the services they did not have time to perform themselves, as well as for entertainment and other urban amenities, as they hopscotched across the landscape following rumor after rumor. The mining West thus broke the familiar pattern of a rural frontier moving slowly and steadily forward, with urbanization lagging well behind the population incursion.

Auraria was not alone for long. In November 1858, William Larimer, an experienced town promoter, arrived in the area. (Town promotion was an accepted, acceptable, and occasionally profitable pioneering venture.) He selected a site across Cherry Creek from Auraria, named it Denver in honor of the governor of Kansas, James Denver, and settled in for the winter. St. Charles fell victim to these newer developments. The sites were nearly the same and the "jumping" (in this case, overlapping city boundaries) had been "greased" by bringing some of the stockholders of the older site into the new company.

Auraria and Denver were rivals for two years, and then merged in April 1860. During that time, the news of gold in the Pike's Peak country was still spreading like wildfire. Some newspapers seemed to compete to top previous stories. For instance, the *Kansas Weekly Press* (October 23, 1858) told of a kettle of gold from Cherry Creek valued at $6,000–7,000. It continued: "Emigrants to the gold diggings have become so common it is useless to ask them where they are bound." They carried the essentials: "washers, pans, picks, wheelbarrows," food, and the all-important "whiskey." Most had no mining experience and must have planned to learn on the job. The *Lawrence Republican* (September 2, 1858) captured the mounting gold fever in verse:

> Oh the Gold!—they say
> 'Tis brighter than the day,
> And now 'tis mine, I'm bound to shine,
> And drive dull care away.

Nevertheless, not everyone was caught up in the gold frenzy. The editor of Brownville's *Nebraska Advertiser* (September 9, 1858) solemnly told his readers: "We advise those taken with 'Pike's Peak' fever to not overdo themselves; we think the disease not dangerous, and [it] will pass off without any serious results, by taking a slight dose of reflection." He further opined that there seemed to be "no intelligence sufficiently reliable" to warrant a "stampede." That same day, the *Kansas Chief* (White Cloud) forecast that many persons "will rake and scrape up all the money they can gather, and proceed to the gold regions, where they will probably meet only disappointment, spend their means, and be left destitute."

These predictions were right on the mark, but the prognosticators might as well have been whispering in a windstorm. Many Americans who had missed the California opportunity a decade before saw a second chance in the midst of hard times. Furthermore, these gold fields were closer at hand

and easier to reach. Thirty days from the Missouri River could put one at Cherry Creek and gold.

As winter set in, little settlements, optimistically called "cities," sprouted up around both the Cherry Creek diggings and along Boulder Creek to the north. With time on their hands, the newly arrived settlers also involved themselves in politics, electing a delegate to Congress and to the Kansas territorial legislature. This did not take place without some dissent, however. One letter writer called it a "partisan, sham election."

As 1859 dawned, the country thrilled to the Pike's Peak excitement, which occasionally managed to push the rapidly intensifying sectional crisis off the front page. In some ways it was 1849 all over again, with the major exception that Pike's Peak country, as mentioned, was located within days of the Missouri River. River towns, both in 1849 and a decade later, promoted themselves as the shortest and fastest way to the gold fields—St. Joseph, Leavenworth, Kansas City, Lawrence, and points in between all trumpeted their advantages.

They were followed closely by merchants hoping to tap the mounting excitement. They offered for sale anything related to that magical phrase *Pike's Peak*: gold washers, gold augers, medicine chests, bacon (which, they promised, would not go rancid during a would-be miner's western adventures), saddles, mules, chewing tobacco, pans, picks, even "Pike's Peak Life Insurance" for the faint of heart. Suggestions on what and how much of each item to take seemed to surface in every store that just happened to have the required goods. For depression-locked businesses and merchants, the frenzy proved a godsend.

Guidebooks promised to get the fifty-niners to their prizes in near "comfort and ease"—as it turned out, too easily and comfortably, really. Miners had a choice of nearly forty such guides, ranging from those that offered bare details to those that tapped the fanciful imaginations of numerous authors (a few of whom had been with the Russell and Lawrence parties or had traveled to Pike's Peak and returned in the fall of 1858). These books were published by printers from the Missouri Valley to New England, areas hit hard by the depression.

Despite the heritage of the earlier Georgia rush, and Southerners' involvement in the discoveries of the summer of 1858, folks in the southern states did not seem caught up in the newest enthusiasm. For them, slavery and states' rights issues were more pressing than a new gold rush.

Their Yankee neighbors suffered no such compunctions. The guidebooks all presented a starting point and detailed a chosen route (with maps, watering places, fuel sources, and camping spots outlined), and some included a variety of other tips, such as supplies needed, mining methods, and stock-handling methods to prevent loss or injury. D. C. Oakes told his readers they should plan to outfit for six months, taking only "what is likely to be needed." That totaled precisely $517.25, and included oxen, a wagon, blankets, gold pans, provisions, matches, a frying pan, and knives, forks, and spoons. Oakes's estimate—more realistic than many—was well out of the reach of many, so they stayed home and merely dreamed of chasing riches.

Some guidebooks honestly described the problems of the journey ahead, whereas others made the trip seem more like a Sunday afternoon picnic. All, however, either hinted at, or bluntly stated, that gold would be found: "buckets" worth, "purest gold," "found everywhere," "veins yielding $3,000," and similar fanciful proclamations limited only by the author's imagination or greed. Lucien J. Eastin's *Emigrants' Guide*, for instance, included a letter from Auraria promising: "Each day brings to light some new discovery. The shining dust glitters in the sand. Miners are making from five to ten dollars a day. Nuggets have been found worth twenty-five dollars and fifty dollars."

Readers often prepared for a lark, leaving their good sense and inhibitions behind. Yet a guidebook coauthored by William Byers warned:

> In conclusion, we would say to all who go to the mines, especially the young, *Yield not to temptation.* Carry your principles with you; leave not your character at home nor your Bible; you will need them both, and even *grace* from above, to protect you in a community whose god is mammon, who are wild with excitement, and free from family restraints.[5]

Good advice, but how many heeded it?

A Rocky Mountain blizzard's-worth of letters, pamphlets, guidebooks, articles, and other promotional material tempted avid readers that winter and spring. Few were able to separate the truth from fiction; not that many cared, with golden visions dancing in their heads. Anywhere they looked, they were bound to find something that they wanted or wished to believe about the Pike's Peak country. They saw little hope, and certainly no such grand expectations, in their stay-at-home neighborhoods, as depressed as the economy was. There seemed scores of reasons to go and few to stay home.

There were a few pessimists among the multitude. Commenting on the guidebooks, the editor of the Kansas City *Western Journal of Commerce* (June 12, 1859) noted: "They have changed the course of rivers, removed mountains, lengthened streams and made bleak hills and barren sand wastes smooth, and even highways."[6] The *Boston Evening Transcript* (May 19, 1859) was even more blunt: "Pikes Peak is a humbug." The editor believed that people were being lured by "the cruelty and atrociously false stories concocted by persons in border towns."

There was more than a bit of truth in both observations, but that was not what people wanted to read. Comments in New York and Chicago newspapers in 1859 were much more to their liking:

> HO for Pike's Peak! There is soon to be an immense migration to the new El Dorado. Many go to dig, perhaps quite as many to speculate on the presumed necessities, or fancies, or vices, of the diggers.
>
> *The New York Tribune* (January 29)

> A bigger army than [that with which] Napoleon conquered half of Europe is already equipping itself for this western march to despoil the plains of their gold. The vanguard has already passed the Rubicon, if I may so metamorphose the muddy Missouri.
>
> *The Chicago Press and Tribune* (February 4)

Naive, gullible, desperate, optimistic, confident—a host of adjectives could describe the gold seekers. Perhaps the story of a Council Bluffs, Iowa "Dutchman," which may or may not be apocryphal, captured it all. Ovando Hollister, in *The Mines of Colorado*, recounted the story.

> A Dutchman [German] . . . was observed gathering up a large lot of meal-bags. He was asked what he was going to do with them. "Fill them with gold at Pike's Peak," he replied. O! he could never do that, they said. "Yes I will," returned he, "if I have to stay there till fall."

Hollister, who knew better by the time he wrote the book in 1866–1867, concluded. "It is [a] matter of congratulation that people have become well cured of such charming infatuation." Not so in 1859: Dreams of gold filled many heads as the new year dawned. So far, the only real footing for all this excitement was the small amount of gold found by the Russell party. However, few questioned the basis of the sensational commotion and increasingly frenzied and exaggerated reports coming east from "gold country."

Meanwhile, out along Cherry and Boulder Creeks and the South Platte River, the vanguard of the "march" who had already arrived settled in to wait out the winter. They anxiously watched for the snow to start melting so they could rush into the mountains to search for the elusive gold that they were convinced was hidden in the recesses of the canyons and peaks to the west.

Experienced forty-niners understood that gold came down the streams out of the mountains. Their technique, learned in California, was to work their way up the creeks and waterways and pan for gold to find the source of the mineral, where it washed into the stream. *There* would be the bonanza.

Snow came early in October of 1858, along with cold nights, but that did not stop those idling in the makeshift settlements along the foothills from looking west and even fighting their way into the mountains. Georgian George Jackson, in January, was hunting and prospecting along Clear Creek (then called Vasquez Creek) when he found gold near the future Idaho Springs. Jackson well understood that one did not brag about a gold discovery, so, as he said, he kept his mouth as tight as a "number 2 beaver trap."

Still, he eventually told a few friends. They formed a prospecting party and struggled up the narrow canyon during the spring runoff. They were rewarded in May 1859 by finding gold at Chicago Bar. In mining terminology, a *bar* is a placer deposit in the slack portion of a stream where the heavy gold, washed out of the mountainsides, accumulates. Finding these bars had been the goal in '49, and it was again during the Pike's Peak rush.

In Boulder, a January prospecting party found gold-bearing quartz high in the foothills between two canyons. Gold Hill they called it, and they knew enough to mind their own business; plus, they were distant from the settlements in and around the future Denver. This discovery, however, was the first to become generally known. William Byers, in his first issue of the *Rocky Mountain News* (April 23, 1859), carried three reports referring to Gold Hill and its vicinity.

By far the most important discovery was made by the Georgian John H. Gregory—another experienced miner, albeit somewhat lazy and inattentive. He too worked his way into the mountains, in April of 1859. He and four companions eventually fought their way up the snow-clogged north fork of Clear Creek to near its head, where they panned and found the bonanza they had been seeking. Some pans yielded as much as $8 worth of gold, rivaling the balmy days of California in 1848 and early 1849. Gregory's strike area later became Central City and Gilpin County, which quickly grew into one of Colorado's greatest gold districts. Like so many other discoverers, however, Gregory did not stay around long enough to prosper from his good fortune. He soon sold his claims and drifted out of Colorado history.

These three discoveries laid a solid basis for the Pike's Peak rush—a rush that had started with the Russell party's findings and accounts that greatly magnified their significance. Americans felt that they had finally come into some good luck. Not only could Midwesterners and Easterners rush to Pike's

Peak, the region's best-known geographical point, but Californians could also stampede over the Sierra Nevadas to the silver discoveries that became Nevada's astonishingly wealthy, famous, and productive Comstock lode. In a unique coincidence of United States history, two major rushes at the same time drew folks hither and yon searching for their personal El Dorados. America's mineral cup overflowed, and for those obsessed with gold, pressing sectional issues receded into the background.

The fifty-niners knew nothing of these three strikes as they prepared to venture westward. For well over a decade, people had been traveling over the Mormon and Oregon Trails along the Platte, and for a generation over the Santa Fe Trail. There were also new and relatively untested routes along the Republican and Smoky Hill Trails. Before the exodus concluded, some would be dead, others shocked by the hardships, and most disappointed that gold did not appear in "buckets galore." Again personal accounts and newspaper reports captured the drama:

> Our streets during the week have been lined with white topped wagons. The vast majority are gold seekers for western Nebraska and Kansas.
>
> *Nebraska City News* (March 12, 1859)

> It is astonishing how rapidly we learn geography. Indeed, ninety-nine out of every one hundred persons in the country did not know that there was such a topographical point as Pike's Peak. Now they hear nothing, dream of nothing but Pike's Peak. It is a magnet to the mountains, toward which everybody and everything is tending. It seems that every man, woman and child, who is going anywhere at all, is moving Pike's Peakward.
>
> *Evening News* (St. Louis) (March 17, 1859)[7]

They journeyed in both conventional and unconventional ways. One emigrant started with six dogs pulling his "light wagon." Another had "two large" dogs pulling his. A projected wind wagon schooner, designed to sail over the "prairie ocean" pushed by gentle westward breezes, sounded great in theory but sank into the first gulch it encountered. Some pushed handcarts, others planned to walk. If they lacked personal knowledge, they depended on their guidebooks, which they hoped contained a good map and an honest account of what they would face. Before it was over, though, some became lost, wandering in the wilderness, and at least one party resorted to cannibalism.

The editor of the *Wyoming Telescope* (April 9, 1859) became concerned about a "number of young ladies" en route to the diggings: "They have little idea of the hardship they may have to undergo during the journey." They probably journeyed on, undaunted, along with 100,000 equally excited fellow Americans. "Pike's Peak or Bust" was their slogan.

They were crossing what had been described by the expedition of Stephen Long as the "Great American Desert." Indeed, for many who were accustomed to water and trees, the land appeared desolate and inhospitable, except along the rivers. Newspaperman Albert Richardson described the land he traveled over by stagecoach in June: "We are still on the desert with its soil white with alkali, its stunted shrubs, withered grass, and brackish waters often poisonous to both cattle and men."

Along the way, the seekers encountered problems they never expected. In a letter published in the *Daily Missouri Republican* (May 29, 1859), an unidentified young man told his father about his experiences. His party had "suffered much on the journey," having taken the Smoky Hill route and been "sorry for it." After the wagon team gave out a hundred miles from their goal, they had to "pack" the rest of the way: "We soon got out of all our provisions, save a few crackers. On these we subsisted for six days, our daily allowance being two crackers each." He concluded, "The suffering on this route has been terrible." On the whole, it was far better to take an established route than to try to find shortcuts over to the gold fields.

Human nature created almost as many problems as Mother Nature. Former forty-niner Henry Wickersham described his 1859 experience: "I find human nature has not changed much since my trip to California. Men going to Pike's Peak now quarrel just as much as men did going to California then. We came very near having bloodshed in camp a day or two since."

Another veteran Californian, and soon-to-be Colorado newspaperman under the nom de plume "Sniktau," E. H. N. Patterson, became very sentimental when he stumbled on the grave of W. Probasco of Caryville, Kentucky. Looking at the "tomb I could not resist a feeling of sadness. . . . [H]is bright hopes of golden treasures in the snowy peaks upon which his dying glance may have rested, had been nipped by the chill fingers of Death."

Not everyone was melancholy, discouraged, or disagreeable. Lawyer Charles Post, who eventually made Colorado his home, said that when they first saw Pike's Peak in the dim distance, his party gave that all-occasion nineteenth-century cheer, Pike's Peak "three times three. It was a beautiful sight, the rising sun shining brightly on the perpetual snowy camp of these mountains made us all feel quite cool. At the same time [we] were delighted to know that the Auriferous Peak after so long and wearisome a journey, was at last in view."[8]

Unfortunately, most reaped only disappointment in the promised land. The ragtag settlements were not awash in gold; in fact, no one seemed to know where it might be except in the snow-locked canyons and mountains to the west. Some went back home within a few days or weeks, as discouraged now as they had formerly been excited. A few wagons carried placards such as "busted by damn" (more politely, "busted by thunder") or, as one "pilgrim" painted on his wagon. "Oh, Yes! Pike's Peak in Hell and Damn Nation." Others went west to Oregon or California, and still others were simply stranded. A father of a fifty-niner had his son's letter published in the *Daily Missouri Republican* (May 29, 1859): "Times here are hard and dull. There is no gold at Pike's Peak. No man can make ten cents a month. I am out of money, and without a chance to make any." He pleaded with his father to send $125 "to take me back home" where he knew he "could make something. . . . If you don't send me some money, I will starve to death. Send in haste." Guidebook writer D. C. Oakes, hurrying westward, met some disappointed Pike's Peakers returning home. Some threatened him; others buried him in effigy and left variously worded epitaphs:

> Here lies the body of D. C. Oakes,
> killed for aiding this damned hoax.

When the "go-backers," as they were derisively called, departed, and reports started trickling back home and into newspaper offices, the rush gained this new label of "hoax." The *Missouri Republican*'s editor bluntly described the situation, albeit with a Victorian flourish, on May 11:

> Destitute of provisions or means of conveyance, disappointed and utterly disheartened, with broken hope and blasted fortunes, toil-worn, foot-worn, and heart-weary, these wretched adventurers come straggling across the plains in squads of dozens or scores.

Henry Wickersham was not discouraged, however. He stated: "I place no reliance in any of the reports. I want to see for myself. Nearly every man we meet tells a different story." He philosophically concluded, "It seems a man can't travel on this road two hundred yards without forgetting how to tell the truth."[9]

Among those racing west was William Byers, who planned to start a newspaper, the *Rocky Mountain News,* in Denver. He was not alone: John Merrick had the same idea with his *Cherry Creek Pioneer*. Both Byers and Merrick knew the importance (and potential profit) of owning the first newspaper to hit the streets of Denver. Byers won the race; the *News* beat out its rival with a first edition late in the evening of April 22, although it was dated the next day. The *Pioneer* lost the race by a mere twenty minutes, according to unofficial timekeepers, but its initial issue was also its last. Merrick was the first, but not the last, newspaperman, or merchant, or businessperson, to find out that in mining rushes, it pays to get there early.

The *News* gave Denver City a definitive advantage in the struggle with its urban rivals "Golden City," nearer the mountains at the mouth of Clear Creek, Boulder City; and the soon-to-be Central City. The popular tag "city" implied something about the community to the outsider, though none of these struggling communities could yet live up to the claim. Having a newspaper to promote and advertise a city and what it offered was one of the best hopes for success. Both the publishers and the would-be cities knew, though, that all their efforts would be for naught if the "hoax" image and the reverse migration were not stopped and replaced with reliable news. Only the really gullible would be fooled a second time. At that point, two men, almost single-handedly, turned the region's future around: William Byers and America's best-known newspaperman, *The New York Tribune*'s Horace Greeley. Byers was already on the scene, and so had a strongly vested interest in the outcome. Greeley, however, did not.

Greeley and Albert Richardson, a young journalist, were on a trip to Utah and California to give their readers reports about the rapidly changing West. In early June, they stopped in Denver, which was rife with stories of the three solid strikes, to inspect the new town and get the real story of the rush. The area where Gregory and his friends were mining appeared to be the best-known and richest, so Greeley and Richardson decided to visit Gregory's diggings. They were joined by correspondent Henry Villard of the *Cincinnati Commercial,* who was also visiting Denver.

With Byers's encouragement, Greeley, Richardson, and Villard journeyed to "see the elephant" in Gregory's Diggings. Greeley found out at first hand what mountain travel was like, including struggling over "steep, rugged" trails and being caught in a "smart shower, with thunder and lightning." Crossing a stream, his mule stumbled, and his fall left him stiff and sore by nightfall. Nevertheless, the party reached the diggings the next day, and found the miners ready for them.

They encouraged Greeley to try panning, which he did—and found some gold, as they had known he would. Greeley also "visited during the day a majority of those which have sluices already in operation." That evening, June 8, a mass meeting, lit by burning pine branches, was held, primarily to discuss the organization of mining districts and the institution of a set of mining laws. Before that, however, their famous visitor was called upon to give a speech to the reported 2,000 to 3,000 assembled miners. "Three cheers" greeted him. Not one to turn down such an opportunity, Greeley promptly admonished the miners about the temptations of gambling and drinking, encouraging them to maintain good order and "to live as the loved ones they left at home . . . would wish."

Those admonitions may or may not have fallen on listening ears. His support for "the formation of a new state" and praise for the "vast future before this region" caught their attention, however. After three rousing cheers, several other speakers took the platform, touting the "flattering prospects of the mines and the rich treasures in the gulches and ravines." Richardson hit upon two popular themes: "The late discoveries promised to add a new star to the federal constellation, and to locate the great Pacific Railroad of the future in this central region." The eastern visitors interlaced a few jokes about "mules and mule-riding" before the speech making concluded.[10]

Greeley recounted his adventures in *An Overland Journey*. Leaving Denver, his party started up a hill, a "giddy precipice."

> Our mules, unused to such work, were visibly appalled by it; at first they resisted every effort to force them up, even by zigzags. I was lame and had to ride, much to my mule's intense disgust. He was stubborn, but strong, and in time bore me safely to the summit.

A lot of fifty-niners and others had stories to recount about mules. They, and burros, became legendary in the mining West.

For his Victorian stay-at-home readers, fascinated by the goings-on in the West, Greeley recounted his adventures in a report from "Gregory's Diggings, June 9, 1859." They learned that six weeks earlier, the "ravine was a solitude, the favorite haunt of the elk, the deer, and other shy denizens of the profoundest wildernesses." By the time of publication, though, "probably" one hundred log cabins were being constructed, "while three or four hundred more are in immediate contemplation."

> As yet, the entire population of the valley—which cannot number less than four thousand including five white women and seven squaws living with white men—sleep in tents or under booths of pine boughs, cooking and eating in the open air. I doubt that there is as yet a table or chair in these diggings, eating being done around cloth spread on the ground, while each one sits or reclines on mother earth. The food, like that of the Plains, is restricted to a few staples—pork, hot bread, beans and coffee.

Greeley guessed that "less than half of the four or five thousand people now in this ravine have been here a week; he who has been here three weeks is regarded as quite an old settler."

Greeley also gave some sound advice: "And I feel certain that, while some—perhaps many—will realize their dreams of wealth here, a far greater number will expend their scanty means, tax their powers of endurance, and then leave, soured, heartsick, spirit-broken." He himself adhered to "my long-settled conviction that, next to outright and indisputable gambling, the hardest (though sometimes the quickest) way to obtain gold is to mine for it." He maintained that "a good farmer or mechanic will usually make money faster, and of course immeasurably easier, by sticking to his own business than by deserting it for gold-digging."

The insightful Greeley understood the significance of activity besides mining that was taking place in the diggings. "Mining quickens almost every department of useful industry." A blacksmith was on the scene, sharpening picks "at fifty cents each," and a "volunteer post office [had been] established." Looking into the near future, he foresaw that a provisions store would soon be needed, "then groceries, then dry goods, then a hotel, etc., until within ten years the tourist of the continent will be whirled up these diggings" over easier roads. This visitor "will sip his chocolate and read his New York Paper—not yet five days old—at the Gregory House, in utter unconsciousness that this region was wrested from the elk and the mountain sheep so recently as 1859."[11]

Greeley and his party went back to Denver the next day, and remained there for several weeks. His June 20th dispatch made a more sober assessment of the gold craze, returning to his concern about whether all this would pay off: "I answer—it will pay some; it will fail to pay others. . . . but ten will come out here for gold for every one who carries back so much as he left home with." Victorian rhetoric overtaking him, he concluded that "[t]housands who hasten hither flushed with hope and ambition will lay down to their long rest beneath the shadows of the mountains, with only the wind-swept pines to sigh their requiem."

He warned his readers that the distance from the settlements, the elevation, the high cost of almost "every necessary of life," and the need for capital to develop the mines made this anything but a "poor man's diggings." He could not have been more blunt in proclaiming, "this is not the country for you!" Repeating his earlier theme, he wrote, "Far better to seek wealth further east through growing wheat, or corn, or cattle or by any kind of manual labor, than to come here to dig gold."[12]

Optimistic and positive, yet constructive and moralistic, Greeley gave his readers a remarkably insightful, vigorous account of the early days of Colorado mining. He saw a great future for the West, and confirmed that the gold was no mirage. The perceptive Greeley also realized that what the region needed most was a railroad built all the way to the Pacific Coast. His Pike's Peak readers could not have agreed more.

9

1859: The Year Dreams Became Reality

When the speeches ended, the distinguished "keynoter" Greeley became an onlooker as the miners set about creating a rudimentary set of mining laws. Technically, those mining on Cheyenne, Arapaho, and Ute land were trespassers, who had no legal rights to their claims. Their company also included those who were mining in parts of the Kansas, Utah, and Nebraska territories. All these very interested parties gathered at a miners' meeting, a democratic institution that brought everyone together to chart the development of a "district."

They had a precedent: Californians had taken similar measures when they overran the Sierra Nevadas a decade ago. Veteran miners knew the legal course to take, and they blended U.S., Mexican, German, and Spanish precedents to create their rules. This was truly functional American grassroots democracy, although the indigenous inhabitants of the land were notably absent from the table. The proceedings gave the miners some internally legitimate basis for ownership, but they were still trespassing—a matter of little concern to most, who believed strongly in manifest destiny and did not care who stood in their way.

First, they defined the Gregory District and clarified their intentions for the area. Everyone, whether experienced miner or excited "pilgrim," would have an equal opportunity to make his fortune, with the district's original discoverer being allowed to stake an extra claim. Twelve resolutions were approved. They were simple, democratic, to the point, and involved few officials—perfect for fifty-niners who would much rather be working their claims. The resolutions included:

- "No miner shall hold more than one claim except by purchase or discovery."
- Each miner could hold one mountain claim, one gulch claim, and one creek claim "for the purpose of washing, the first to be 100 feet long and fifty feet wide, the second 100 feet up and down the river or gulch and extending from bank to bank."
- Mountain claims must be "worked within ten days from the time they are staked off, otherwise forfeited."
- "Each discovery claim shall be marked as such, and shall be safely held whether worked or not."
- A "company constituted of two or more, shall be at work on one claim of the company, the rest shall be considered as worked by putting a notice of the same on the claim."
- "When two parties wish to use water on the same stream or ravine for quartz mining purposes, no person shall be entitled to the use of more than one half of the water."

To resolve disputed claims, the district secretary was to list nine disinterested miners, from which list each party could strike off three names. The remaining referees "shall at once proceed to hear and try the case, and should any miner refuse to obey such decision," a miners' meeting would be called. If the decision by that body was the same, "the party refusing to obey shall not be entitled to hold another claim in this district" and shall pay the secretary and referees "$5.00 each for their services."[1]

Once the excitement of the Gregory District's opening meeting subsided, other proposals soon appeared regarding "needed" modifications. Unsurprisingly, some suggestions proved to be incompatible; others were so convoluted as to be nearly incomprehensible. The simple solution was to appoint a committee to look into the issues. At a July 16 meeting, the committee reported that the original code should be retained, albeit with some additions.

A recorder would keep all district records, including mining and water claims and bills of sale. Also, a one-dollar fee would be paid to the recorder for each claim recorded. All laws relating to trials of disputed claims were repealed and new ones promulgated. To accommodate the fast-changing methods of mining, quartz mill and tunnel sites were brought under district jurisdiction, as were companies that brought in water. They could pass over any "claim, road, or ditch" but could not "injure the party over whose ground they pass."

The Gregory miners held no further meetings until November, at which point they fine-tuned what they had already accomplished and made a few modifications to meet some new need. Thereafter, they focused their full attention on the reason they had come to the Pike's Peak country: mining.

This was only the beginning. Before they were finished, Gilpin County had been divided into nearly a score of mining districts, each with its own laws. All the districts in Gilpin County established rules and regulations through the same process, although some interesting content variations appeared.

Russell and South Boulder Districts gave women the same rights as men in staking claims, but the stubborn men in the Silver Lake District emphatically declared (in the secretary's original spelling), "be it annacted that no woman shal hold claims." The Russell and Fairfield Districts stated that every person "of suitable age" residing in the district "is hereby declared a voter."

Miners in the Pleasant Valley District carefully defined another problem that repeatedly caused trouble: "No miner shall run tailings or throw waste dirt or rock upon the adjoining claims without permission of the owners except such as will naturally run in the water from Tom, Rocker or Sluice."[2] Violators were liable for damages.

Some speculators planning to make money in "urban" real estate appeared on the scene. The Hawk Eye District allowed such ventures, but required a plat; it allowed the developer only "every tenth lot." The South Boulder District permitted towns to be laid out provided "the person or company" gained "the consent and signature of a majority of the miners in and of this district."

Miners in the Climax District would not permit "Liquor selling or [any] Gambling Establishment" and granted the constable "twenty five cents per mile going and returning for all necessary distances traveled serving papers." Both the Independent and Russell Districts prohibited "houses of ill fame or prostitution." The rest remained silent on the matter, thereby tacitly allowing the world's oldest profession to flourish. On April 28, 1860, Russell voters categorically stated that they wanted to advance "the interests and promote peace harmony order and a good understanding between man & man and believing that the allowing Counternancing or encouraging of low Body Houses Grog Shops and gamboling Saloons to be degrading to the Morals detrimental to the sway of peace and order and Disgraceful to the name and character of the District." This resolution also levied a $50 fine on a "party or parities" violating the regulation.

In this loose-knit society that often became litigious over mining issues, lawyers might have had an unfair advantage over the unrepresented or inexperienced. The Independent District bluntly took care of that potential problem: "No practicing lawyer, or any other person having been admitted as such in any State or Territory, shall be permitted to appear in any case pending in the District, as attorney or agent of any person." A lawyer could appear only if "a legal party to any case" or if both parties employed counsel.

In a virtually rootless society, jails and prison terms cost tax money to build and oversee, none of which the fifty-niners wanted to pay. Consequently, they devised other, more expedient means of punishing infractions. The Illinois Central District, for example, approved of fines, lashes on the "bare back" (from 10 to 100, depending on the crime), and banishment. "[A]ny person found guilty of murder shall be hung by the neck until dead" or, interestingly, shall be "banished from the mines and property confiscated." A jury decided which penalty to apply.[3]

Experienced miners in all districts realized that water was critical to their ventures. As mentioned earlier, the Gregory miners' meeting discussed the issue on June 8. After the dry summer of 1860, they tackled another issue in February 1861: "in case there shall not be water Sufficient for all, priority of Claim Shall determine the right to such water." The Independent District carefully defined the water power granted to run mills and the responsibilities of water companies; the Wisconsin District defined water claims and ruled that they "shall hold as real Estate, & not [be] jumpable." Miners of the Pleasant Valley District were concerned about obstructing the free passage of water and building dams that backed water onto another's claim. The secretary of the Bay State District might not have been much of a speller, but readers grasped the meaning:

> It Shall be the privilege of enny miner or Miners to take out the water out of North Clear Creek in a ditch or floom around anny mans Claim

or over his Claim for the purpose of washing Dirt on the Hill Side by hydraulic power or Slusing not ingering the claims passing thare over.[4]

In future Boulder and Clear Creek Counties, miners had been equally busy. With an extra-legal basis in place for owning land, miners could now develop their placers and their mines, mill men could build their mills, promoters could develop towns, and water ditch plans could be implemented. Rudimentary government had been developed and everyone could focus on the real reason they had come to the Rocky Mountains. A pattern had been established that would carry over to every new discovery area.

All this was probably secondary to those who had rushed to Pike's Peak country for one reason and one reason only: gold. Reports going east still varied from the truthful to the dismal to the exciting. The report by Villard, Richardson, and Greeley carefully explained: "Gold-mining is a business which eminently requires of Its votaries capital, experience, energy endurance, and in which the highest qualities do not always commend success." They decried the "infatuation" with gold that had caused "thousands to rush to this region" and then "run back before reaching it." A letter writer quoted in *The New York Tribune* (May 28) concluded that at "present there is very little here to induce people to come out, yet they keep coming." Indeed they did, causing the *Atchison Union* of June 4 to theatrically complain that the rush had placed "thousands, and tens of thousands in irretrievable ruin."

Writing to his nieces from Mountain City on July 17, L. D. Crandall explained what underlay such reports. While on his way west the previous month, he had met people who claimed "Pike's Peak a perfect failure & a humbug & every story to discourage coming here that human imagination could invent."[5] William Green Russell knew better, even though by the time he returned the Gregory diggings were so overrun with prospectors that he moved southward into what became known as Russell Gulch. He reported to his hometown newspaper, the *Dahlonega Mountain Signal* (August 19), that "new discoveries are being made every day" and mountain diggings were improving. He also stated that many gulches and ravines "are yet very little worked or prospected."

One of the region's prime boosters, William Byers of the *Rocky Mountain News*, visited the Gregory diggings in mid-June. He wrote, enthusiastically, that "every day witnesses new gold digging discoveries." He expressed a "wish" that would echo down through Colorado mining for decades, and elsewhere as well: that the "mines already opened grow richer as they descend." Byers reported that about 100 sluices were operating, that the "product safely set down" was $200 per day per sluice, and that pans paid up to $10.[6]

The Gregory diggings, by far the richest of the three initial discoveries, received the most attention during the late spring and early summer of 1859. There Colorado mining earned its initial fame—or infamy. In this district the miners learned their first real mining lessons. Gold appears in nature in the form of free gold, found in streams, or as lode (vein) gold, combined with other minerals and *gangue* (waste rock) that had to be mined out of the mountainside. The fifty-niners quickly encountered both in the Gregory diggings. Fortunately, they initially started by placer-mining free gold.

The pan came in handy, both for gathering their golden treasure and when sampling up a stream to find where the ore was eroding off the mountainside. When gold no longer appeared in the pan, the prospector knew it had to come from below that point. None of this took very long to learn, but panning was backbreaking work.

Like the forty-niners before them, these miners wanted to make their fortunes quickly. To make more money, they had to move more "dirt" (gold-bearing gravel). Hence, they built rockers. After ore was shoveled over the rocker's sieve, miners poured water over the ore, then rocked the whole mechanism back and forth. The idea was that the heavy gold would settle on the bottom and get trapped behind the cleats, while water took out the loose gravel and sand. Essentially, this was what happened with panning, too; the point of the rocker was to wash more dirt faster and a bit more easily.

The simplest methods quickly spawned innovations and equipment developments. The first improvement was the sluice box, a long wooden trough stretching out twenty-five to fifty feet, with cleats on the bottom and another sieve-like device at the top. After ore was shoveled in, water carried it down the sluice, allowing the gold to settle behind the cleats.

Fearing they might still be losing gold, the miners transformed the sluice into a long tom. Pushed by water, just as in nature, gold settled to the bottom behind numerous cleats. It took several men, however, to run such operations, and so mining companies and partnerships emerged. These methods also used more water and demanded more land on which to dump waste rock. The water flowing back into the streams carried sand, dirt, and whatever else came out of the sluice and the long tom. Soon men downstream protested vociferously—hence the initial district regulations on this subject.

No matter what process was used, final cleanup depended on a vital fact: mercury's affinity for gold. When the two were mixed together, the mercury picked up the precious metal from the black sand (containing minerals, usually iron, nearly as heavy as gold) that generally was left behind with the free gold. Careful heating separated the gold from the mercury, leaving the miner with his treasure and with expensive mercury that could be reused—as long as he avoided breathing any of the poisonous mercury fumes.

Once placer gold was located, prospectors usually tried to track it back to its source. Even while the streams were busily being mined, the fifty-niners started digging gold out of the mountainsides. To do this successfully meant leaping into industrial mining. Placer skills were easily learned, but following a lode into the ground required different skills, techniques, and equipment.

To ensure that the mountain did not collapse on them, the miners had to timber the shafts, drifts, and tunnels they dug. Seeping and pooling water had to be pumped out. Ore had to be brought to the surface, and that meant hoisting it up the shaft, which required some form of power. The hoisting mechanisms evolved swiftly: initially, manpower operated a windlass, then a horse powered a whim, and finally steam powered a mechanical hoist. Unless the owner and his partners operated the mine themselves, miners had to be hired to dig the ore, and the vein had to be followed carefully to increase profitable production and keep expenses to a minimum. To protect equipment and provide storage, buildings were required. All told, mining was expensive and time-consuming.

Once the ore reached the surface, the gold had to be separated from the gangue. At first, miners used the straightforward old Spanish / Mexican *arrastra,* which involved grinding the ore in a circular stone basin or trough by dragging a large rock over it. The miner then introduced mercury into the pulverized ore and recovered his gold.

These basins were easy to build and easy to operate, but were not fast enough or able to handle a large enough volume for those in a hurry to get rich and get home. The obvious answer was the stamp mill, which had been used for years in other gold districts. This mill used a steam engine to drive a battery, or batteries, of heavy iron "stamps" (usually five to a battery) to pound the ore. Mercury introduced into the finely crushed ore bound with the gold, and the heating / refining process already discussed freed the gold.

Stamp mills also proved fairly easy to build and operate, but they did not always capture a high percentage of the gold; sometimes the miners lost half (or even more) of their treasure. Further, the mills were not initially available locally in the Pike's Peak area and had to be imported, generally from the Midwest. This added cost and time before they could be placed in service. For most, a trial-and-error period also delayed successful operation.

During those tumultuous summer days of 1859 in the Gregory diggings, the opportunity for "poor man's diggings" had not yet waned, despite the amazing pace of activity there. Frank Fossett later tried to capture the excitement:

> Before the end of May the valleys of the streams that course through the mountains of the country that has since become Clear Creek, Gilpin, and Boulder counties were alive with men. Trees were felled, cabins erected, and sluice boxes constructed for washing the gold from the gravel and "pay dirt." Hand rockers were also used and arrastras were subsequently operated.

Another early chronicler of this era, Ovando Hollister, noted that "[b]y the lst of July, 1859, there were one hundred of these sluices running within a short distance of Gregory Point." Because water, for both the gulches and lodes, was scarce, two ditches of considerable magnitude were projected, to bring it from the head of Fall River. These eventually became "The Consolidated Ditch . . . [which] cost one hundred thousand dollars, and for the soft gristle of the young community was a big undertaking and quite a triumph" (it was finished in the spring of 1860).[7]

Reports from the mountains varied as summer slipped away. Nevada Gulch reported "large nuggets" and "old mines" operating steadily. In the Gregory diggings, however, water was getting scarce, which led to work being "temporarily abandoned on many good paying claims."

L. D. Crandall's letters reveal how one man balanced dreams and practical reality. After a few weeks in and around Mountain City, he was convinced "that there was and is large quantities of gold in these mountains but I am further satisfied that it costs more labor & [is] harder to begot at than in California." Still, he purchased four claims and determinedly wrote: "Gold is here and I shall stay here until I get some of it and will stay here through the winter and next season"—a clear demonstration of "gold fever" in action.

With so many working the streams, the productivity of the placers started to decline, spurring some of the crowd to rush off to any newly discovered "pound diggings"—rich areas yielding three to twenty dollars per pan. Those who stayed had to turn to the serious matter of digging underground. The *News* encouraged them, reporting that quartz mills with several steam engines were on their way from the States (July 23). On September 10, the paper noted a "large quartz mill" being erected in the Gregory District. "Before long," the paper predicted, people would see the "result of systematic quartz mining."

It turned out that establishing a mill was difficult, too. The first steam quartz mill in the United States, erected by Prosser, Conklin & Co., went into operation on September 17. It had a successful run until late in the month, when "huge columns of smoke and steam" were seen coming from the mill. The fire caused a total loss of more than $10,000. In October, the Coleman, Lefebvre & Co. steam mill started operating, but soon broke down. It was repaired and started again in November, but that was late in the mining season. Others were still coming or planned when winter closed in. Industrialization had not yet overtaken the solo prospector, nor underground mining the search for easy placer pickings.

George Jackson's discovery of gold in South Clear Creek, in January 1859, was not publicized until April, and then was soon overshadowed by Gregory's discovery on North Clear Creek. Nearly all the miners rushed northward, but attention soon returned to South Clear Creek's placer deposits. Among those who arrived in the Clear Creek diggings at Payne's Bar was a young couple, Augusta and Horace Tabor. Like many others, they had been struggling on a Kansas farm when the news of the Pike's Peak discoveries reached them in 1858. They determined to go to the gold fields to improve their financial situation. They started west in the spring of 1859.

Leaving his wife at the future Golden, Horace struggled into the mountains, was too late for the Gregory diggings, and tried his luck at Payne's Bar. As he wrote later, "I located a placer there and worked on it all summer and made moderate wages." Augusta was more revealing. She took up the story when her husband and two friends returned to Augusta's camp:

> They thought they had better move me up farther. We packed up and went beyond there up to Payne's Bar. We were three weeks going from there to where Central is now. Had to make our road as we went. We could only make about three miles a day, a wagon had never been there before.
>
> There miners told Mr. Tabor he ought not to keep me at Idaho Springs during the winter on account of snow slides that would cover us all up. He became frightened and moved me back to Denver, when he returned to camp he found his claim had been jumped. Some of the miners had told him this to get him away so they could jump his claim. There was no law in those days.[8]

The Tabors had learned a valuable lesson about abandoning a claim and about human nature. They were not alone in making mistakes; the inexperienced and the gullible needed to be continually on guard and careful about "friends."

Others were more fortunate in mining the bars along Clear Creek that summer. As in the Gregory District, water shortages hampered operations until the late summer rains came. The *Rocky Mountain News* reported on October 13 that the Clear Creek mines had been nearly abandoned for the season, particularly those around Idaho Springs and above.

What eventually became the richest of the Clear Creek discoveries that season was also one of the slowest to be developed. Two brothers, George and David Griffith, were among those who arrived too late at the Gregory diggings and turned their attention elsewhere, heading over the mountain to Spanish Bar. Working their way upstream, they eventually reached the future Georgetown area by August. There they found gold-bearing quartz outcroppings on the side of a mountain that they named for themselves. Joined by their brother John, the three men sank a shaft on the Griffith Lode and worked several other promising claims. Then, lured by news of rich strikes ("thousand dollar" diggings, some said) to the north, they left the valley temporarily. (The Griffiths eventually returned. That fall, one stayed in the area and the other two went back east, one to purchase supplies and the other to bring his family back the next year.)

When William Byers visited the Georgetown mines on August 27, the Griffiths were gone, but a few other miners had arrived. Typically, Byers was impressed:

> Two or three other claims are being opened on the lead, but not yet properly worked, although they evidently are very rich. Several other leads crop out in the immediate vicinity, on both sides of the creek presenting more favorable opportunities for working them . . . but there is no one to develop them; the whole settlement consisting of five men and one boy.
>
> The creek is also rich in gold, but the bed rock is deep and it requires a combination of labor or capital to work it successfully.

Despite that warning, Byers enthusiastically told his readers (September 10): "We have no doubt there are millions of gold in the immediate vicinity."

On Fall River—a tributary of South Clear Creek—quartz outcroppings were found, ditches dug, and "slides" constructed to get the quartz down from the hillsides. Despite much work and time spent, disappointment in lieu of profits led to the mines' being generally abandoned by the fall of 1859. Of the three initial discoveries, the Gold Hill District received the least attention, even though it was the first one generally known, thanks to Byers's report in the first issue of the *News*. Hal Sayre, just starting his Colorado career, humorously described his experience in this district. He and a partner arrived, neither having had any mining experience or even having been in a gold-producing region.

> There was plenty of gold in the vicinity of Gold Hill, but Peabody and I failed to find it. We ran over the ground almost like fox hounds seeking a trail, and nearly as rapidly, stopping to stick a pick here and there, but without any knowledge of indications as to where to go to work. We found other parties making the search much as we were doing.

The same could have been said of fifty-niners in every mining district.

By July of 1859, both placer and quartz mining were well under way in many districts, but information seemed difficult for Byers to obtain. It was, the *News* reported, on July 23, "understood" that miners were making between three and twenty dollars per day. (Those same familiar figures popped up in every district.) The article concluded, "on the whole we report progress very encouraging and have still stronger faith in next year's operations." The enthusiastic Byers never stopped promoting!

The so-called Twelve Mile diggings, twelve miles from Boulder, and the Seventeen Mile diggings both had their moments that summer and fall. In late October, however, the *News* reported that although the Twelve Mile diggings had had "several hundred men this past summer," at the present time "not more than fifteen men were working, most making three to five dollars per day."

A former Missouri slave was one who seemed to be doing well. He showed the reporter three dollars in gold he had taken out of his rocker in half a day, and his claim appeared "very rich." Others were having less success. A "rude" arrastra on Left Hand Creek proved to be a "failure" that discouraged quartz mining. The reporter thought it far inferior to those worked by either oxen or water power in the Gregory diggings.

One report, published in September, concluded that the quartz rock taken out "cannot be advantageously worked without machinery." The next year would offer better prospects with the arrival of needed equipment. An October 6th letter pointed out that the "quartz leads" were "very numerous and very rich" and that there "are some very rich gulch diggings" near Gold Hill and at Twelve and Seventeen Mile gulches.

Even less success was achieved in the Jefferson and Deadwood diggings, "about fifteen miles south" of Gold Hill. The miners had spent much time making miles of ditches and opening the creek bed to bedrock. Unfortunately, few "indeed, were crowned with the success which they deserved for their perseverance."

Farther afield, August found excitement way up north along the Cache la Poudre River. On August 13, the *News* reported "rich diggings" and another "will-of-the-wisp" stampede. Among those who ventured north was John Gregory, looking for another strike. It seemed that wherever they looked now, prospectors envisioned—and occasionally found—the gold they had dreamed of when they came to Pike's Peak country.

Far more exciting than even the Gold Hill diggings was the news that gold had been discovered in South Park. Miners who found the "older" districts overcrowded had started moving further into the mountains by midsummer. They entered mountain-rimmed, wind-blown Bayou Salado, better known as South Park. They rushed to Tarryall, but tarried only briefly before finding their "bonanzas" in early August at what became Fairplay and in the mountains nearby. The *Rocky Mountain News* (August 13) bore by-now-familiar headlines: "Rich Discoveries," "Great Stampede." Reports of "$100–$1,000 per day" for the "lucky miners" soon appeared, but the *News* warned that they were "not entirely well authenticated." The rush, the paper said, became a "continual stream of miners" with wagons, carts, and pack animals. "What the result will be, time alone will tell, but we believe rich mines will be found along the Platte throughout the whole extent of its mountain course."

A familiar pattern was repeated. The editor told his readers on September 17 that "hundreds of men rushed expecting to gather gold by the thousands." Disappointed, they cried "humbug" and "rushed away again as precipitously as they went in." He advanced the usual advice: "Those who went patiently and industriously to work are now making money at the rate of twenty to sixty dollars per day to the man."

Putting aside their fears and worries, the fifty-niners crossed the "snowy range" and landed on the Western Slope of the Continental Divide. The first prospecting party failed to find the fabled "pound diggings." Worried about the Utes on whose land they were trespassing, they retreated. Others, who crossed over the range later in the season, discovered gold dust in paying quantities in several gulches.

They scattered along the Blue River, placering and tracing veins up the gulches and mountainsides. A letter appeared in the *News* (October 20) as the mining season drew to a close. The correspondent had been there "near two months," found the stream bed to be very rich, and had "no hesitancy in crying out Eureka." Another letter writer opined that the "mines prove the most extensive yet and regular paying diggings of any yet discovered in the Rocky Mountains." Such enthusiasm could not overcome the fact that they were terribly isolated in a country that was itself isolated. Still, these pioneers stayed, built cabins, and started a little settlement that would one day be called Breckenridge.

In the midst of all the gold excitement, the *News* (August 13) carried a small headline "Silver Mines Discovered," followed by a brief article. Twelve miles above Boulder, a "silver mine was recently discovered" and "believed very rich." On August 20, the *News* expanded its story, saying that a former slave with twenty years' mining experience had discovered the "reputed silver mine." Whatever the miners found, they claimed they traced the vein for two miles. The *Rocky Mountain News* printed an August 22 editorial stating that silver had been found "in many places of almost incredible richness and unlimited extent. We believe the Washoe and Arizona will be outdone." That was some belief, because the Comstock (Washoe) was still glorying its first bonanza. In the September 12 issue, the newspaper claimed, "[D]iscoveries of silver veins still continue to be made almost daily in all parts of the country. As might be naturally expected the whole country is on the tiptoe of excitement respecting silver mines."

Apparently, the Griffiths also found silver but did nothing about it. All this did was stir up a minor tempest in a teapot over who had discovered silver first and where it had first been found. Such a strike was part and parcel of the legends that led some people to expect that they would uncover rubies, diamonds, and other precious stones at any moment. At the time, though, gold overshadowed everything else. As the summer turned to fall, nothing more was heard about silver mines. More importantly for the future of the territory, coal and limestone were "found in large quantities" near Boulder. The mining industry and the camps in general would need both if settlements were to become permanent.

Wherever they prospected and mined, the fifty-niners faced medical and health problems, ranging from bad drinking water to accidental shootings to mining accidents. With few physicians within reach—and some of those poorly trained—and modern medicine nearly a century away, the miners, often poorly fed, badly clothed, inadequately housed, and working long hours in wet conditions, fell victim to a host of maladies.

The altitude affected some, particularly those who rushed high up into the mountains where exposure to the elements might be deadly. Bad water, bad food, and bad whiskey did in others. Many complained of rheumatism, blaming the altitude and the dampness of mines. For some, nearby hot springs provided at least momentary relief, if not a cure. Thus was born Colorado's image as a "health mecca." According to the imaginative promoters, the springs would relieve almost all known aliments, and they summoned the ill and infirm to bathe in or drink the "miracle waters"—a call that expanded especially after the railroad arrived.

In the summer of 1859, a mysterious "mountain fever" ran through the mines in Gilpin County and elsewhere. Although it was never medically identified, many thought it was a typhoid-type disease. It proved debilitating and sometimes deadly. Whether it was one disease or several will never be known, but Henry Villard described it as a "ravaging disease" that "demanded many victims, and caused many more to abandon their work and seek the plains." Perhaps some Midwesterners, coming from the fever-ridden districts in their home states, had carried latent fevers with them.[9]

In contrast, the out-of-doors work proved beneficial to some fifty-niners. L. D. Crandall commented that his health had "been generally better than it was in the states. The worst of my case is being obliged to go with wet feet the most of the time."

Caught between dreams and discouragement, a constant smaller rush of go-backers ebbed into Denver and points eastward. By late summer, it was becoming obvious that the weather would soon change, especially in the mountains. The downward flow increased as people scurried out of the mountains with the coming of the first snow. A Mountain City correspondent wrote to the *News* (September 12) that "many of our miners are now preparing to leave for the States for home." He was convinced, though, that "all of [them] express their determination to return early in the spring with their families and to make future homes among us."

Among those who departed was John Gregory, whose gold discovery had been the prime factor in saving the Pike's Peak rush. Byers understood that Gregory had almost single-handedly saved the day and hailed him: "No one had labored harder to develop our resources than he nor with better success." He took with him the "good wishes of the whole country" and, probably more importantly, carried "$25,000 in gold dust" back to his home in Georgia, after previously forwarding "some $5,000 to his family." Byers hoped to see him again in "the coming spring with machinery and facilities such as we know he will operate as it should be done." When the spring of 1860 came, Gregory arrived with his quartz mill, which he soon sold. That was his last hurrah. After drifting around for another two or three years, John Gregory disappeared from Colorado.

Meanwhile, as stream flow decreased in late 1859, the number of sluices operating declined, and people started journeying back to the states, or at least lower elevations. The first snows hastened the process for some, but this was the first year in the mountains for many, and they did not know how quickly the snow could melt on warm fall days. Wrote one *News* correspondent after it snowed three to four inches on September 17, some "talked of suspending operations"; then "old sol came out and melted the snow." On the 29th, twelve inches fell, "naturally causing serious alarm," and "some set out immediately for the valley." Then sunshine came out again and in five days the snow was gone. By late October, the outward emigration was nearly over. The *News* reported that those remaining "generally feel in good spirits, all looking forward with the brightest hope to the next mining season when they expect to make their pile and be among the homeward bound."

Meanwhile, busy Byers continued to refute negative stories about the mines and the region. In his November 10 edition, he pointedly called a defamatory Illinois article a "base fabrication of falsehoods." Earlier that fall, he became so carried away that he refuted a story point by point and called the author a "d—m fool"—strong language for the young newspaper.

As 1859 wound down to its final weeks, snow finally stayed and winter took hold. It had been an amazing year and not just for mining, which was well on its way to becoming industrialized. Some enthusiastic politicians had created and organized the territory of Jefferson, not only to provide another legal basis for the mines and towns but also, no doubt, to create some paying jobs. Roads had been built, trails improved, towns and camps established, better connections made with the "States," homes and buildings constructed, and the essentials of Victorian society rooted. Some light industry could be found in Denver, the largest settlement and transportation hub. Thanks to Byers and his *Rocky Mountain News*, Denver's future looked bright. With only one paper in the territory for a brief period, all the news filtered through it. His outreach to the states back east praised his town, its prospects, and its investment and settlement possibilities.

Agriculture, both farms and ranches, was found along the creeks and rivers tumbling out of the mountains. Some fifty-niners, tired of or disgusted with mining, returned to their Midwestern farming roots, with Byers's encouragement. He loved to print the "amazing results" of their efforts in bushels per acre and size of vegetables. Byers looked to the time when the region would be self-sufficient, although that dream was slow in becoming reality despite his best promotion. His well-taken point, however, was that the Pike's Peak country would be vulnerable until locals could produce more of their own food and grains. The cost of living remained high, and everyone knew why: the need to ship goods across the plains (and the high freight charges to do so). This fact of life was starkly highlighted as winter storms closed or hampered that exposed, essential lifeline.

Another threat to that lifeline existed year-round. Plains tribes, as mentioned earlier, had been upset for several decades about white men's incursions on their land. They understood clearly what was happening to their way of life and their culture. A Plains Indian war could severely hamper and perhaps even damage development prospects for years. The Utes in the mountains faced the same predicament, but so far had caused only scattered problems.

The region had been thoroughly explored along almost the entire foothills area and well into the mountains. Prospectors and miners had penetrated South Park and crossed the mountains into the Blue River Basin. Isolated camps and mines stood as outposts in rugged land, each convinced it was the future El Dorado.

These future Coloradans might be isolated by distance and the winter season, but they could not escape what was happening in the country they had left. Sectional tension had increased alarmingly that fall following the attack by radical abolitionist John Brown at Harper's Ferry, [West] Virginia, and his dramatic trial and execution. A presidential election would be held in 1860, with prospects bright for the young Republican Party to take control of the White House and perhaps Congress. "Black-hearted abolitionists," Southerners called them, and threatened secession if the once-unthinkable occurred; that is, if Northerners elected an anti-slavery president. Heated talk on both sides included mutterings of war if that happened.

Some of the fifty-niners had fled the states in a forlorn hope of escaping the mounting tension and name-calling, but found that they could not outrun the failure of American democracy. Moreover, their local newspaper kept them aware of the changing situation, as did letters and newspapers from back home.

Nevertheless, the fifty-niners accomplished a huge amount in a very short time. That speedy development showed the essence of the fast-moving, urban mining West. Not everyone had been successful, as Byers admitted on July 23. "It is true that many very many are still unfortunate." Still, others had prospered enough to go home with a "stake" they could not have obtained so quickly in any other way. Many more hoped to do the same the next year. Optimism, confidence, grit—these folks had it.

One of Byers's correspondents explained, in an October 27 letter, that he was "confident that there is a fortune for each, lying hid in these mountains, which they mean by labor and perseverance, to obtain, and they are little concerned whether others believe it or not." For him and his determined contemporaries, the new mining season could not come soon enough. Another wrote in September that "old gold miners" who had worked in California, Australia, and South America "all concur that the Gregory district, for six miles square is richer in gold than any other of the same extent in the known world."

The fifty-niners had faith. They were not go-backers, and some even thought this new land was not a bad place to put down roots. Maybe even a few of them could still hear the faint echoes of the Cherry Creek emigrants' song that had cheered them that past spring: "There's plenty of gold, in the west we are told in the new Eldorado." Whatever they might have thought about gold prospects as 1860 opened, a new mining season was nearly upon them.

3

1860–1864: "To Everything There Is a Season"

> Last Monday Rob, Sam and I started out to look around and in the afternoon we discovered a lead which we called the Moline and have been at work on it ever since but can not tell whether it is going to turnout to by any account or not but we have good indications so everyone tells us that know any thing about it but if it is not worth anything we intend to "try and try again" for I believe that there is plenty of gold here and if any one will just pitch in with heart and hand they will get some of it.[1]

William A. Crawford, writing to his cousin Mina from the Rocky Mountains, captured perfectly the expectations of "Young America," as the press liked to call the generation that opened the second season of Pike's Peak mining. William and his crew were living in a tent "on the mountain about one half of a mile back of Gregory Point and we are enjoying ourselves very well. The only fault we have to find is that we can not hear from home oftener"—a "fault" that was the subject of many and frequent complaints. This determined worker criticized the go-backers who, when they found they "cannot pick the gold up by the handful[,] . . . whine around for a few days and then sell out and start for the states," but also gave some familiar advice: "I think that those that have farms and families and are doing well had better stay there."

The rush of 1860 never equaled the one of the year before, but Ovando Hollister declared that "never did spring open on a more hopeful people. The richness and great extent of the new gold fields were considered proven." By the first of May, "Immigrants were arriv[ing] from the States at the rate of a hundred a day."

What made Hollister's outlook so sanguine? Among other things, the mining experience that had been gained, the mining machinery that had arrived, the laws and regulations that had been established, the roads that had been built and bettered, and the support industries and agriculture that had been developed. Finally, he noted a "great influx of people, many of them women, many of them bringing more or less capital, and coming at least prepared for the emergencies of the season."[2]

Such boosterism could not overcome the fact that the Pike's Peak mines were not newsworthy back in the states. That story was now "stale"; other headlines grabbed readers' attention, including those on the increasing sectional hatred, the fading hope of some resolution before a national tragedy ensued, and the presidential nomination conventions, campaign, and election of Republican Abraham Lincoln. All these pushed Pike's Peak and golden fantasies into the background. Then, South Carolina's secession destroyed all hopes for peace.

Some people went west that spring to escape the impending war; others came for the same reasons the fifty-niners had. Much of the Pike's Peak country had not been prospected, and it was hoped that the known districts had not been completely stripped either. As Hollister noted, "In view of the unparalleled richness of the lodes and gulches in the Gregory District, the most sanguine expectations were not unreasonable." Reality proved different, however.

William Crawford, for example, did not have the success he had forecasted in 1860. He wrote to Mina on September 2, 1860, from Mountain City, where he had decided to stay for the winter, even though some of his friends were returning home: "I think that I may stand a chance of making something more than wages by staying here for we have several claims that I think will see for something next spring."

Next spring, Mina received a letter saying that William was looking for work and planned on going prospecting. Like others, he confided in his

March 31, 1861, letter why he had not returned home: "I think that there is no use of making a baby of myself and running home just because I feel like it when I know that I could do better here, or, think so at any rate." In October of 1861, Crawford said he had planned to go to Gold Hill, but gave up the idea and stayed at "Bobtail City" near the Bobtail Lode below Central City. The optimism he had cherished since the spring of 1860 had vanished.

Philosophically, he noted the high prices of potatoes ($2.00 to $2.50 per bushel): "that seems to you like an awful price, but it appears cheap to us." Although he had maintained good health, others had been less fortunate. William commented, "I believe this would be a very healthy country if people would live as they do in the states."

William had no luck in 1861 either. In May of 1862, he told Mina that he had not much news to write, "for I am still working by day" on the "No. 6 on the Bobtail" Lode. "There are a great many idle," he complained, whose unemployment woes were heightened by "high prices of grub [and] of all kinds of provisions." Being a veteran miner now, he looked with skepticism on the Salmon River (Idaho) excitement: "A great many have star[t]ed there from around here." Showing that he was not completely isolated from the rest of the world, Crawford noted that he was glad to see "the war going on with so much spirit and energy."[3]

Crawford's letters tell us much about the times, from health to high prices of goods. Both his optimism and his lack of success were typical; nevertheless, "next spring" always held promise. Although he wanted to become a mine owner, nothing worked out for him and he ended up a working miner, resolved not to go home until he had made his "stake." Thousands like him would continue their quests in Colorado during the high tide of mining.

The spring of 1860 was not so wonderful for newcomers, either. "Old-timers," with all of a year's experience, did not welcome the immigrants who crowded into Boulder, Clear Creek, and Gilpin Counties. Hollister explained what happened and why:

> They felt that they had earned what they had got, and that there was chance enough for others to do likewise. Surely, they said, all these strangers cannot expect employment here on our ground; let them branch out and find mines for themselves, or if not, go back.[4]

The newcomers generally turned to prospecting, "which is a discouraging business except to the prospector by nature, who must have the faith of a martyr," or they made "continual purchases of claims which they knew not how to work." Then Mother Nature chipped in with a heavy snowstorm in early May, "well calculated to dampen the spirits of the new men."

Finding pickings slim in the established strike areas, the newcomers spread out seeking the new El Dorado that surely lay just over the next mountain or along the next stream. With the Eastern Slope districts already overcrowded and claims staked, the hopefuls ventured onto the Western Slope. The Blue River diggings promised less competition, and what lay beyond them sparked the imagination and drew the adventuresome even further westward.

Again, however, the mining tide ebbed and flowed. As Hollister noted, miners found "wages were quite low; it seemed there were no more big things to strike; it was hard to get letters from home; and is it wonderful they felt discouraged?" Hollister continued: "By the first of June [many] might be seen facing homeward with a dejected air, as if under conviction for sin. Still, the incoming tide was vastly the strongest." He estimated that during a two-month period, at least 5,000 per week arrived. "They generally brought grub enough to live on for a few months, they wandered around, joined every stampede—of which there were many—in short, didn't think of returning to the States till fall."[5]

Prospectors moved across the range throughout the year—Georgia, Humbug, Galena, and French Gulches all had their moments. Gold Hill, Gregory, and Russell Gulches proved that the "old-timers" were not exhausted, particularly after a reliable water supply arrived for Gilpin County thanks to the Consolidated Ditch. Mining in Clear Creek County evolved quickly during the year: "Placer mining had been chiefly abandoned ere the season was half gone, and all the energies of the people bestowed on quartz-vein discoveries and development."

Two significant discoveries were made during the mining season of 1860: California Gulch and the San Juans. Each became a storied legend in Colorado mining. At the time, though, the former offered the year's best hope.

A prospecting party left Russell Gulch in March 1860 and eventually reached the upper Arkansas River Valley, after struggling through deep snow. They were not the first; others had been there the year before, but left when winter arrived. On or around April 26, 1860, one Abe Lee, while panning, reportedly shouted to his friend, "By god, I've got California in this here pan." Thus the gulch was named California, and on April 27 the newcomers organized a district.

A "flying rumor [of] $20-$25 each man per day" excited nearby prospecting parties, who rushed to the strike. Perhaps 5,000 had arrived by summer and as many as 10,000 by the end of the year. California Gulch was *the* big mining event of the year, promising to be the spring of 1859 all over again. Lewis Dow wrote to the *Rocky Mountain News* (May 30, 1860) that thanks to the "pleasant weather," the snow, which "had been the principal drawback, is now mostly gone." He went on describe what had become the typical aspects of a mining rush:

> People are constantly coming in from other diggings as well as from the States. Merchants, blacksmiths, shoemakers and attorneys are sticking up shingles here. At present there is a great scarcity of provisions [and those they had were selling at high prices].
>
> The mines opening are richer than even the most sanguine hope for. On May 12 Earl, Hopkins & company washed one small rocker from 150 buckets of dirt [that brought] the sum of $147.20 . . . . Such news is apt to cause a stampede. I advise every man doing well in any other place to let well enough alone. All the claims are taken.

A little mining camp named Oro City soon sprawled along the lower end of California Gulch, and that was where the Tabors again tried their luck. Augusta continued their story, again revealing much about the times:

> We arrived in California Gulch May 8, 1860 and [by] 1861 we had acquired what we considered quite a little fortune about $7,000 in money.
>
> I was the first woman in California Gulch. We killed our cattle that we drove in and divided it among [the others in the camp]. [They] built me a cabin of green logs and had it finished in two days.
>
> I never saw a country [1859–60] settled up with such greenhorns as Colorado. They were all young men from 18 to 30. I was there a good many years before we saw a man with grey hair.

As she had the previous year, she "commenced taking boarders with nothing to feed them except poor beef and dried apples." Horace mined and the couple eventually opened a store. Meanwhile, Augusta was appointed postmistress. "With my many duties, the days passed quickly."

As winter settled in, mining stopped. A conservative production estimate for Oro City places the value at between $2 and $3 million, but no accurate figures are known. Many miners were reluctant to reveal their gold yield, and placer gold could easily be taken out of the district and territory. The production figures also vary according to a subject of constant upset for the miners: low prices for their gold. Whether in California Gulch or elsewhere, the price of gold was set based on its presumed purity, not on the price it would bring back east—and the folks in the Colorado placer districts could do little about that.

Given the disappointments in 1860, hopes mounted for 1861. In April, it seemed to be happening: Iowa Gulch, two miles south of Oro City, grabbed attention, but did not pan out to be another bonanza, despite its "brilliant prospects." With the snow gone, miners returned to California Gulch, which began "to assume the appearance of olden times," with diggings there "being worked with success." By late August, however, a dismal report said that California Gulch "looks a good deal like a graveyard, for several miles along its upper end." The writer went on: "Most folks are talking about leaving, they don't know where; but generally speaking, the gulch is 'dried up,' and the gold is not flush here any more."[6]

California Gulch's placer-mining days had ended. A familiar pattern emerged: In Colorado, placer deposits proved rich but shallow. If the territory was to remain a viable mining region, the future rested with hard-rock mining.

These let-downs were somewhat countered by the revival of another tantalizing story: silver. Ovando Hollister told of silver being found in 1861 in the future Ten Mile District west of Breckenridge, over the mountain from Oro City. The district was out of the way and still undeveloped when he wrote in 1867. On October 10, 1860, the *News* ran another story about "scores of [silver] lodes" having been discovered in California Gulch and elsewhere. The author's veracity remains in question, though, as he went on to describe "opal discoveries" too.

At Buckskin Joe reports of silver seemed to be springing up everywhere. Byers was not convinced. He observed that a "country like this must always have some prevailing excitement, something that for the time being obscures everything else." Just now, the territory "is in the midst of a silver mine agitation" (*News,* August 29, 1860).

Concerned about this agitation, he returned to the topic two weeks later (September 12): "As might naturally be expected the whole country is on the tiptoe of excitement respecting silver mines. Gold mines are deserted, and their heretofore busy occupants busily hunting for places in which to quarry out blocks of silver." He did not want to squelch enthusiasm completely, but he concluded that although there was no longer any doubt that "there were good silver mines, we shall expect many of the so called leads to turn out to be nothing."

Once again, notwithstanding the cautions and bitter experience, the rush continued. Hints of bonanzas spread by that mysterious, always buzzing mining-camp grapevine. The "gold belt" extended from Central City through Oro City and beyond and "in all probability [it] grows richer as one goes in that direction." Prospectors had reached an even more isolated location far to the southwest in the remote, jagged, towering San Juan Mountains, where the Spanish had prospected and mined nearly a century before and the Utes currently held sway. Such was the lure of gold that a prospecting party ventured to New Mexico, turned northward, and eventually followed the Animas River into a mountain-ringed valley later named after the leader of the expedition, Charles Baker. He mined for a season and then retreated before the oncoming winter.

Baker actually did more promoting than mining. He spent the late fall and winter writing letters about his discovery of a second California. His campaign worked: in the spring of 1861 came the "ho for the San Juans" cry—off to the newest El Dorado. As many as 500 ventured that far from the rest of Colorado, and some came northward from New Mexico. They found that Baker had conveniently laid out a toll road into the park and a little village, Animas City, to the south in the lower Animas Valley. He thus charged them for yielding to the temptation of his writing campaign, and also for getting in on the ground floor of an urban boom. Both projects promised money for him.

William Byers called Baker's find a "humbug" and advised others to "go slow." Some had doubts, but the lure of gold proved too great. Unfortunately, the eager seekers at Baker's Park found little gold. Some suggested hanging Baker, but the entire fiasco ended by midsummer as men struggled to get out as fast as they had arrived. Some nearly starved before reaching settlements, and others ended up much the worse for wear.

The Utes' displeasure about the trespassers, the struggle required just to get to the region, the small amount of gold found, the outbreak of the Civil War, and the oncoming winter all combined to end Baker's dream and leave the San Juans with a bad reputation. The two rushes in 1860—California Gulch and the San Juans—display the best and the worst of human nature; the constant allure of gold; and a pattern that would be repeated, to some degree, throughout the next generation.

The valuation issue continued to bedevil the miners, as it had the year before. Gold might be valued at $20.67 an ounce, but that was for pure, refined gold, which was seldom, if ever, found in nature. The fineness of the gold determined the price the miner received. The figure ran from $14 to $18, or around $16 per ounce on average. The ore also carried a bit of silver.

This situation, understandably, generated deep unhappiness and repeated complaints because merchants and "gold brokers" based their prices on the going local rate in the district. Resentment exploded when miners, upset with the $14 to $16 price range, refused to "deal with any man" who would not pay $18. These problems were not resolved until Clark, Gruber and Company opened a private mint in Denver in 1860. The company legally produced $5, $10, and $20 gold pieces. To make certain its coins would be favored for circulation, the company put just a little more gold in each one than federal coins contained. Two other mints operated briefly but achieved neither the success nor the reputation of Clark, Gruber.

The miners were not above a bit of dirty dealing themselves. Some would mix a little extra foreign material into their gold dust; brass shavings were particularly popular. A sly merchant might spill a tiny bit of gold dust when weighing it for a purchase, or put a little grease on his fingers when he took the popular "pinch," which was valued at 25 cents. All told, both parties had legitimate complaints about crookedness on the other side.

Meanwhile, as Oro City and the San Juans rose and fell, other news grabbed headlines. As imminent war roiled the eastern states in early1861, the miners far away in Pike's Peak country received a long-awaited prize: territorial status. On February 28, President James Buchanan signed the bill authorizing the Colorado Territory. The southern states, which had long opposed the creation of another "free territory," had recently marched out of the Union, opening the way for a territorial birth certificate. At last, the miners—and everyone else—had a legal basis for land ownership, a legal and recognized government, help from Uncle Sam, and the hope of someday joining the Union as a state.

A government had already been developed for Colorado when war broke out in April. For the next four years, territory inhabitants lived in the shadow of that epic, tragic event. Coloradans were far away from the front, except for a few months in 1862 when a Confederate force marched into neighboring New Mexico and drove up the Rio Grande. They were eventually defeated with the help of Colorado volunteers, who went south in March to fight the battles of Valverde and Glorieta Pass. Although Colorado did not entirely escape the turbulent times, despite its distance, Coloradans' attention generally remained on their own problems.

Finding a method to extract gold became the overriding problem of the 1860s. Gold dug from the earth was not in a free form, of course. That did not matter for the first fifty or so feet, because of "secondary enrichment"—"nature's way of making high-grade ore out of low-grade ore."

The simplest explanation was that freezing, thawing, and water percolation had, over eons, taken the sulphur out of what was called "decomposed rock." This natural process allowed the miner's mercury to easily free the gold. Below about the fifty-foot level, however, the ore contained sulphides, or what Central City folk called "sulphurets." Sulphides interfered with the mercury refining process, and thus most of the gold could not be extracted and was "lost." Ovando Hollister tried to explain this vexing mystery to Coloradans:

> Somehow the mills as a general thing do not save the gold; why it is hard to tell. There must be some difficulty beyond the mills—doubtless the want of experience in the men who run them. Crushing quartz is a new business to them; and as it is a very nice, requiring skill, and as all of us as yet lack this skill, we fail in almost every attempt, just as any one does in a new vocation, about which he knows nothing.

Ores that had assayed out at hundreds of dollars per ton returned, when milled, only seven dollars or so, which barely gave the miners any profit after deducting the milling expenses. That precipitous drop in yield alarmed everyone, mystified many, and led to a decade of experimentation that met with failure after failure. At first, attempts centered on building larger mills and finding purer "quicksilver," but this avenue of attack was soon abandoned.

Stamp mills, the earlier answer, carried three problems into the new decade. At best, they recovered only about three-fourths of the gold actually present in the ore processed; when the many inefficient operators and faulty equipment were added to the equation, the average production was alarmingly low. Also, whatever silver and copper the ore might have contained (granted, not a high percentage) disappeared into the tailing pile or downstream. The most significant problem was the collapse of the gold recovery rate once the miners and refiners had to deal with sulphide-bearing ore.

In a very real way, the territory's future rested on resolution of the milling and smelting problems. Successful and profitable methods had to be found if Colorado mining was to continue. Thus, as the war dragged on back east, mining trends continued to evolve. Hollister summarized the situation:

> During the years 1861 and '62 placer-mining in Colorado saw its palmist day. It was gone about systematically and not feverishly, and with more experience, generally paid better than it had previously. . . . In 1863 the business gradually died away, because the gulches, most of them worked over three or four times, had become completely exhausted.

Writing from Nevada City, Samuel Mallory put his finger on another cause:

> But I am satisfied that many leave for want of "pluck" and lack of energy, while others get awful homesick, and are ready to cry wolf when there is no wolf, for the sake of an excuse to go home.
>
> A large number come here with less than a cent in their pockets—when here are too lazy or constitutionally tired to do work. I am decidedly of the opinion the latter class have no business in these mountains.[7]

A new wrinkle appeared with the introduction of hydraulic mining, which had been developed in California and was put to use in Colorado. In this process, water, shot through a hose under pressure, washed gravel off a hillside and sent it through a long tom—after which gold was recovered in the traditional way. It cut costs by reducing the number of miners needed (proponents claimed that two men could do the work of thirty), speeding up the entire process, and—it was hoped—enabling profitable working of lower-grade gold deposits.

The early hydraulics operations, along the Blue River down from Breckenridge, left behind quite a heritage. Not everyone was callous about the environmental impact of mining. When the editor of Tarryall's *Miner's Record* visited Georgia and Humbug Gulches (September 1861), he penned this account: "They are tearing up 'Mother Earth' at a fearful rate, undermining houses, and removing every obstacle to their progress in searching for the precious ore."

"Hydraulicing" continued, though it faced problems other than environmental degradation. A July 1863 letter from Gold Run explained that the "drawback to full development" of immense riches was the "scarcity of water." Also, "the great scarcity of hydraulic material such as hose and pipes" presented a further obstacle. The writer pleaded with Denver merchants to send what they had: "We miners must have it, let it cost what it may." Concluding on an upbeat note, the writer pointed out: "Miners are beginning to see the advantage of mining with hydraulics over all other methods."[8] Advantageous or not, hydraulic operations never flourished in Colorado, primarily because of lack of water—plus the fact that the gulches had already been worked over three to four times, leaving little gold available even for those mining in large volumes. Basically, placer mining had seen its day in Colorado.

Even newer placer deposits faced the usual ups and downs. Soon after its June 1860 discovery, Three Nation Gulch (named for the discoverers—an "Irishman, German and Negro") disappeared from the news. A report from French Gulch in July 1861 described it bluntly—"everything dull"—and pleaded for more capital to "develop resources." At one time or another, that plea was heard from every Colorado placer and hard-rock district. A month later, a story from neighboring gulches, such as America, Galena, Georgia, and French, reported "many miners doing well."

Despite the lack of easy pickings, however, Summit County climbed to near the top of the production charts during those years, outproducing the pioneer Gilpin, Boulder, and Clear Creek Counties. Still, even in Summit people "left the failing gulch and placer diggings," according to Frank Fossett. The *News's* certainly reflected that fall-off, as fewer and fewer stories appeared about Breckenridge and the Blue River diggings.

From these districts, the peripatetic prospectors continued probing in every likely (and even unlikely) spot. Some called it a romantic occupation, but a diary that the *Rocky Mountain News* published on April 26, 1862, about an August 1861 adventure, should have given its readers more than momentary pause. The anonymous author started by observing. "Everyone knows that a prospector's life is one of hardship, intrigue and unknown exposure."

> August 11 Rain, rain; bedclothes have been wet a week.
>
> August 12 Much fallen timber though made headway under many difficulties. After traveling three days had to back track. Came to a canon could not get through.
>
> August 20 Very wet traveling Boots thoroughly saturated with water, pants ditto.
>
> August 21 Had fifty [?] of cake for breakfast and dinner. Lost a small piece of bacon yesterday so have none to eat. At noon picked mess of pepper-grass, whortleberry and strawberry leaves. Boiled them in salt and water. Enjoyed vegetable dinner together with bread and coffee.
>
> August 23 No food missed the right trail went over eight miles on wrong one raining

Finally, on August 25, the author reached a ranch, civilization, and food.

Mining ebbed and flowed in the early 1860s. Spring Gulch near Central City was "very nearly washed out[,] as evident from its deserted appearance," observed a *News* correspondent in August 1861. A few miles away, at Peck Gulch, W. R. Campbell proudly wrote to the *News* (January 30, 1862) that claims "are for work, not for sale. We do not go on the 'humbug puffing' sort of principle."

Meanwhile, Gunnison County made a brief bow on the mining stage. For a while, Washington Gulch became the promised new El Dorado. Fifty miles west of California Gulch as the crow flies—a distance doubled by the round-about mountainous overland route—the region equaled the maligned, discredited, and deserted San Juan mines for isolation. Still, Denver's *Colorado Republican* waxed enthusiastic about the area's prospects: Miners "are doing well and everything warrants belief that these mines will prove the best of any yet discovered in this territory" (November 16, 1861). However, the "immense emigration in the spring" that the paper predicted never came about. The remote Gunnison County lay well within the territory of the increasingly unhappy Utes.

Times did not improve for the struggling territory in 1862, as its main economic pillar teetered. Prospectors were stampeding to the future Idaho in 1862 and then to Montana in 1863. It did not help when rival Nevada claimed that the "Colorado gold mines are about abandoned." That brought forth a spirited rejoinder from Central City's *Colorado Mining Life* (November 1, 1862): "Why Lord bless, bless your soul, people are coming in thick every day. Abandoned! Bah! People don't 'suck' that one." Jealousy between mining regions and even neighboring districts never languished far from the surface.

Letters came down from the mountains saying "Pike's Peak has not 'played out.'" Byers jumped on this, telling his readers on July 19, 1862, that "Colorado will yet come out all right." He did repeat an earlier warning: "This is not the country for the poor man to become suddenly rich in, though there have been such instances." The editor of Central's *Tri-Weekly Miners' Register* (November 10, 1862), positively gushed, however, proclaiming that everyone in Clear Creek and Gilpin Counties "seemed satisfied with the summers' work and consider Colorado the paradise of America."

Mining in "paradise"? Not all the news proved that cheerful, nor as hopeful as the reports of previous years. For instance:

> Chicago Bar fair prospects
>
> Spanish Bar seems almost deserted
>
> Georgetown Griffith district not yet well developed
>
> California Gulch scarcity of water, severity of weather—disadvantages
>
> Boulder County suffering from "want of capital and accompanying energy"[9]

The territory had one new excitement in the summer of 1862, when the Tenth Legion Mine burst on the scene. Empire City (Coloradans simply loved to attach that label to every little burg) basked in the moment. The "future Rocky Mountain metropolis" and its Union District grabbed the spotlight in an otherwise depressed scene. "People are flocking in there fast," as both placer and quartz mining developed, aided by natural advantages "which excel those of every other portion of the territory." Though the excitement persisted through the winter, before the war was over, times became "dull" at the never-to-be "metropolis," and its mines offered no more than a "good reputation."

In these otherwise unpromising days, one legend grew up around Central City's Pat Casey, an illiterate Irish immigrant, who parlayed a lucky strike into status as the "celebrated king [or prince] miner of Colorado." The stories of his antics—like the time he once supposedly asked for half his crew to come to the surface when an odd number were working in the mine—grew with each telling and his career, before he went back "east to lose his fortune," gave everyone hope.

The fortunate Caseys have always been a rare species. The more down-to-earth Samuel Leach attempted to explain to his brother George why he stayed in Colorado and what mining meant to him:

> The mining interests me, it is much more exciting than anything else I have ever worked at. There is always the element of chance that fascinates. But it is hard work just the same. You have to stick at it and keep your eyes open. And if you do get a good claim you had better keep still about it for if you get excited and tell your friends someone is sure to hear of it and jump the claim or make trouble for you in some other way.[10]

Thus Leach succinctly captured the essence of "mining fever" and the industry. Few people have better explained what gold fever and mining meant to them and their generation.

Leach and Casey aside, as 1863 dawned, the mining future and fortunes of the young territory looked bleak. New mining frenzies elsewhere shoved Colorado into the background, the mills were not producing miracles at saving the gold, many of the districts were sliding into decline, investment money remained in short supply, and the boom days faded as people chased rumors and dreams elsewhere.

The year 1863 brought no expectations of brighter tomorrows. The "poor man's diggings," as some defined placer mining, generally disappeared. As the mines grew deeper, the cost of everything from hoisting ore to solving water problems mounted. The cost of timber and fuel rose as well, and the lack of skilled miners still plagued even decently producing areas. People were leaving in greater numbers than the previous year for those new discoveries up north. "Bannack on the brain," Denver's *Weekly Commonwealth* (August 6, 1863) chided them, adding that this "bogus Eldorado" has "created great excitement among the uneasy portion of our community." A letter from Breckenridge reveals the new face of gold fever: "Miners don't trust new discoveries as they did a year or two ago."[11] They stayed at home and worked instead.

As always, though, there were some gleams of hope. The world's most skilled miners, Cornwall's "Cousin Jacks," were starting to arrive. Englishman Maurice Morris, visiting Colorado in 1863, noted "the large influx of Cornish miners," who were expected to "improve the practice, if not the theory, of mining here." The Cornishmen complained almost immediately about the "rude way many of the shafts are constructed" and the great "risk consequently run working them."

Lack of money seemed the most pressing issue: money to buy machinery, improve transportation, solve the riddle of refractory ore, and develop the "plentiful" gold deposits. The region, in desperate need of investors with funds, decided to test the eastern market for stock and mine sales. Back east, a war-stimulated economic boom had resulted in profits for many, who began to worry about inflation and the government's turning to paper currency rather than continuing to use gold and silver coins, the "money of the Bible." In short, Easterners were seeking an investment, and what was better than gold? Nothing seemed safer or more dependable as a foundation for one's wealth.

In the late summer and autumn of 1863, the fates of Easterners and gold were joined with those of the future Colorado. Coloradans and their agents hawked their schemes all over the east; Easterners listened and leaped. Suddenly, Colorado properties and companies became the glittering star of the stock market. Every untested claim or hole-in-the-ground mine blossomed overnight, in the minds of a mesmerized public, into a "bonanza" waiting to be developed. Claims were staked in the snow and sold to unsuspecting buyers. New York and Boston investors eagerly devoured the numerous mining pamphlets flooding the market and clamored to buy.

They were told that the mines grew richer with depth, that the gold was inexhaustible, and that the mining problems were solvable. Investors fell all over themselves scrambling to get in on the ground floor. Buy they did—worthless claims, overpriced mines, undeveloped prospects, and nonexistent holes in the ground owned by liars. They sent out untested mining and milling machinery, accompanied by wild expectations, inexperienced mine captains, and piles of money. Coloradans could hardly believe their luck.

Frank Hall, from his office in the *Daily Mining Journal,* cautioned: "Those who have the true interest of Colorado at heart, will uniformly frown on all attempts to swindle parties in the East by selling at enormous figures property which has no known existence, outside of the Recorder's Office." On April 6, 1864, he returned to the subject:

> Never in the history of the country did the excitement of mining speculation run so high as at present. Men are every day receiving windfalls in the shape of receipts from the sales of claims to Eastern capitalists. Demand increases rather than diminishes. . . . This state of things is unhealthy, and therefore can not be permanent.

By late winter of 1863–1864, it became obvious to those not completely bamboozled that something was wrong. The substantial financial outlay from the east was not matched by gold bars coming from the west. The boom came to an end in April 1864. There was no pot of gold at the end of the Colorado rainbow. Faster than it had developed, the gold-mining investment frenzy failed, with cries of anguish and blame from both east and west. The situation amounted to a "slaughter of the innocents," according to *The New York* Times (April 6). "Not one in fifty mining ventures ever pays a profit, and not one in twenty ever returns their investment to the stockholders." Angry name-calling and finger-pointing offered no consolation to those who had lost money and had dreams crushed. They determined not to be burned again by the unprincipled—if not downright criminal—Coloradans. Speculation and greed, combined with desperation and dishonesty, gave Colorado a bad reputation that persisted for more than a decade.[12]

Colorado mining stock lay dead on the market. Investors shied away, reputations were ruined, and the whole sorry mess left a bitter, devastating heritage for the young territory. Further, the sad episode had solved none of the territory's mining problems; in fact, it had made the situation worse. The exhilaration and profit-taking produced only short-lived benefits and left behind long-lived grief.

A young Virginia City newspaper reporter, at Nevada's silver-ribbed Comstock, remarked on the Colorado situation even as he speculated on Comstock "feet," as they called stock. Samuel Clemens (alias Mark Twain) wrote about his mining-stock ventures in his classic book *Roughing It*. The lesson Twain learned was this: "There are two times in a man's life when he should not speculate: when he can afford it, and when he can't."

John Wetherbee, in his 1863 volume on the Colorado Territory, summarized these years succinctly: "Like a lottery, the prizes attract." Many had gambled their futures in that "lottery," but few indeed claimed any prizes as 1864 slipped away.

4

1864–1869: "Good Times a-Comin"—Someday

"It was the best of times, it was the worst of times." Charles Dickens had published those words just as the Pike's Peak rush started. His popular *A Tale of Two Cities* undoubtedly traveled west that summer, packed away in some fifty-niner's luggage, and as 1865 dawned, Coloradans could apply his description to their own situation. They were certainly glad to leave the year 1864 behind; the "gold bubble," as the mining-stock frenzy came to be called, had lifted them up and knocked them down in less than a year, but its effects in the territory lasted for the rest of the decade.

The *Rocky Mountain News* of April 18, 1864, deplored one negative effect of the gold bubble: "The panic in the stock market consequent upon the fall of gold has left Colorado mining property pretty dull. The expenses and delays that have embarrassed the Colorado companies the past season have discouraged many." The ever-optimistic William Byers, however, added that confidence in the richness of the mines of Colorado was "not shaken."

Nevertheless, even Byers was worried. In the December 11, 1864, issue of the *News*, he fretted about there having been so many failures in Colorado mining that "people are now afraid" to invest. The wealth was there, he assured his readers, "but the territory has not fulfilled its glittering promise of two years ago." Despite forecasting a bright future with the new "separation processes" being built, Byers vacillated; eleven days later, he headlined an article, "The Curse of Colorado," pleading with mine owners to stop running other mining properties down "to build up" their own.

The boom days of 1859 had become the depressed days of 1864, followed by a disheartened 1865. Byers advised his readers in the October 6 issue of the *News* that they should not leave Colorado, for "good times were just around the corner." For most of these "migratory, restless men," however, both the present and the future looked bleak.

In addition to the crash of the gold-stock schemes, there were problems with the Plains tribes, which had been increasing in intensity since early 1863. The tribes considered these pioneers trespassers and destroyers of their land and game, and they knew what had happened to the Sioux in Minnesota in 1862. There resentments had exploded into a massacre, and the army, after a brief war, drove the Sioux out on the plains to join their cousins. In retaliation, the Sioux and their allies virtually cut off transportation in the summer of 1863. They did it again in 1864, much to the disgust of angered Coloradans. Wagons, trains, and stagecoaches stopped running; stage stations were destroyed; nearby farmers and ranchers fled to Denver; and the cost of living in Colorado skyrocketed. The tension eventually led to the tragedy of Sand Creek in November 1864, resulting in a rare occasion when the Plains Indians waged war in the winter.

But this was still the Colorado Territory, where encouraging news and rumors refused to die out completely. Silver, that alluring coquette that had teased men for decades in these mountains, finally materialized from fleeting hopes and dreams into a reality. (Something had to happen, with Colorado gold being in such disfavor.) Given the fame of the Comstock lode, and its silver millions and millionaires, some thought silver might hold out the best hope for Colorado.

It happened in the upper Clear Creek Valley, where prospectors had poked around for years. A few promising strikes had been made, and two little settlements—Elizabethtown and Georgetown (which merged in 1866)—snuggled next to each other amid the mountains and trees at valley end. The profitable placers along the lower valley, where some of the original 1859 discoveries had been made, completely overshadowed this lonely district.

Hard-rock mining, not placers, had produced what gold had been found, but it was never enough even to pay operating expenses, let alone yield profits. More than a decade later (January 12, 1878), Georgetown's *Colorado Miner* concisely summarized those days, noting that it "was a sickly and despondent camp for a long time. The stamp mill could not save much gold because there was but little gold in the quartz. Provisions were high. It was all outgo and no income." Late in the summer of 1864, as the aftermath of the gold-stock mania wreaked its havoc, specimens of silver ore were uncovered in and around Georgetown's mountains. A Central City reporter for the *Miners' Register* arrived in the valley in mid-July 1864, just before some rich silver veins were found. He encouragingly discussed the new mill under construction, proclaiming that the lodes "of that locality are very numerous" and that some ores "yield enormous" amounts of silver. The article included one caveat: "The mountains rise abruptly on each side of the stream, and are difficult of ascent." Even that, however, might prove beneficial for tunneling and intersecting mineral veins. The report concluded that "Georgetown offers very superior inducement for prospectors well as capital miners."

Once the news leaked beyond the valley, prospectors started to appear as summer turned into fall. The *Miners' Register* headlined an October 4 article, "SILVER MINING," in which it said: "Silver mining is usually considered much more profitable than gold mining." The writer explained how the ore could be profitably smelted, for "silver ores are complex ores," as everyone soon found out.

Initially, hopes soared as high as the nearby peaks. Assayer Frank Dibbin predicted that silver assays would "give much better average results than those of Nevada and consequently the silver mines of Colorado are superior to those of Nevada." If Georgetown surpassed the Comstock, the territory would be blessed indeed. More immediately, Georgetown held the possibility of being the next "bonanza." As snow ended the mining season, locals had plenty of rich specimens to show, and Cornish miners had arrived. All they needed was capital, better transportation, and machinery. (Sound familiar?)

The next year, as the war ended, Georgetown and silver appeared to be the future of Colorado. Prospectors climbed the mountains, located claims, dug into the rock, organized districts, and lugged high-grade specimens down to the valley for everyone to see and admire. They cited with enthusiasm the old mining adage that the higher the silver strike, the richer the ore—and the prospectors were ranging ever higher. As a 1867 government mineral report stated, "These veins were followed to an altitude previously unknown in mining experience in Colorado." It continued:

> Much excitement was occasioned in Colorado by this discovery [silver], and a large number of prospectors were soon engaged there, making discoveries and preemptions under the liberal laws of the Territory, which gave undisputed possession to discoverers who should have their claims recorded in the county office, after making the developments and improvements required by law.[1]

Suddenly the future shone bright again. With the Comstock fading into borrasca rather than bonanza, Georgetown briefly took center stage—and once again reality struck the promising district. It was the same old story: The ore near the surface was easy to work, but the farther and deeper the miners dug, the more complex the mining became. As with gold, the milling and smelting methods had to deal with "refractory" ores.

Like their neighbors, Georgetown miners awaited the development of a reliable, cost-efficient process that would allow their ore to become profitable by saving a high percentage of gold, silver, or both. Hopes rose briefly when Lorenzo Bowman, who headed "a party of colored prospectors" in the area back in 1864, revealed that he had had experience in Missouri with lead mining and smelting. He helped devise a solution based on heating the ore in furnaces. Sadly, the method failed with ore from deeper mines. Bowman had no better luck with a Scotch hearth smelting operation.

Meanwhile, the country's leading mining journal, the *American Journal of Mining* (*AJM*), kept its readers informed about developments. "That silver interests of Colorado are entitled to no little attention in mining circles is a fact now being rapidly established" ("The Silver Mines of Colorado," February 2, 1867). "Your readers will be justified in drawing the most flattering conclusions as to their future, from the daily growing testimony labor is affording as to their value. To be sure capital is needed . . . . Smelting is the key to success" ("The Silver Mines of Colorado," June 1, 1867). "The silver region is being rapidly developed. Here too is seen the repetition of the old folly, a mill costing $50,000 with no mine to give it ore" (October 3, 1868).

Finally, in an editorial again titled, "The Silver Mines of Colorado," in the April 24, 1869, issue, the editor advised readers that Colorado's silver district likely had a solid basis "upon which in the course of time a prosperous industry will be built up." He wryly noted, though, "But with the usual half shrewd, half sanguine manner of western mining communities, our friends in and about Georgetown are talking of their prospects more largely than the facts will warrant." In fact, many expected too much too soon. The *Rocky Mountain News* gushed, on June 16, 1868: "It is a wonder to us that there is not a rush from every quarter of Colorado at least to the silver region of Georgetown. The lodes there are proving marvelously rich."

The *AJM* editorial warned Coloradans to stop blaming eastern capitalists for their troubles: "We are not such fools as we used to be." The writer advised that it "is necessary to open and work good veins" and "get your ore," and then capital will come. The editor's final comment is telling: "We hope to chronicle during the present year for the territory of Colorado, not a new speculative excitement, but a sound and permanent progress."

An old Mexican proverb cautions, "It takes two years to make a mine." It does take time to develop mines, improve transportation, and find a profitable smelting method or methods. However, impatient Americans, both owners and investors, wanted silver profits immediately. So, Georgetown languished.[2]

In 1865 and the following years, silver "discoveries" kept popping up, always teasing but seldom producing. Some reports came from areas previously reported to have silver, such as the Snake River and Ten Mile regions. Others were new, such as in the valleys above Georgetown and in Boulder County. Rumors—always rumors—placed silver almost everywhere in the territory. Byers boosted silver as he had previously touted gold. The *Rocky Mountain News* (March 24, 1869) tried to persuade readers that the amount of silver produced that month "ought to convince our Eastern friends that there is silver in the Colorado Mines."

By mid-1869, though, gloomy reality settled over the Clear Creek Valley. Placers no longer produced what they once had, and the silver mines awaited the savior of smelting at a reasonable cost. This moved the *Colorado Miner* (July 2) to observe that while passing from Georgetown to Idaho Springs, the traveler would see a "great many costly monuments."

> These "compliments" to the dead will not be found to resemble marble, nor nothing grand and imperishable, but rather will they be found after the pattern of ruined and *ruinous* mills, surrounded by old rusty machinery, and decked with scattered fortunes.

The monuments, the author continued sarcastically, were "erected years ago by men who had no more idea" of what to do "than the man in the moon has of what is on going in the Georgetown police court room."

United States Commissioner of Mining Statistics Rossiter Raymond blamed much of the problem on "eastern capitalists who did not accomplish anything." His 1869 Georgetown report also pointed the finger at "tons of worthless machinery," "large sums . . . paid for undeveloped property," and "large mills built before the mines were able to supply them." Then he took a pointed swipe at "agents, generally incompetent and without experience in the business."[3]

Despite such pessimism toward Colorado silver mining, as the decade turned Georgetown still held the key to success, with most of the mines and the production that had so far been achieved. Nevertheless, everyone awaited a successful smelting operation.

Silver mining districts were not alone in their melancholy situation. Ovando Hollister painted a depressing picture of Oro City, the once "golden star" of 1861. He noted truthfully that "nothing, indeed, can be more deceiving or ephemeral than the feverish prosperity of a placer mining country." Of California Gulch and Oro City, he wrote graphically:

> The relics of former life and business—old boots and clothes, cooking utensils, rude house furniture, tin cans, gold pans, worn-out shovels and picks and the remains of toms, half buried sluices and riffle boxes, dirt-roofed log cabins tumbling down and the country turned inside out and disguised with rubbish of every description, are most disagreeably abundant and suggestive.

His prophetic hope was that "a new and better and more permanent life be founded upon these quartz deposits, thus fulfilling that Scripture which says 'the last shall be first.'"[4]

Hollister could have said the same of several score of once-promising and prospering placer camps scattered along the Blue and other rivers, even as far away as the San Juans. These relics of a "bygone time" spoke volumes about men's hopes, dreams, and disappointments.

The *American Journal of Mining* again joined in the 1869 condemnation, noting that a hole "20–30 feet in the ground" does not make a mine. With the Georgetown folk, they blamed eastern capitalists for letting those properties remain unworked or only marginally active. All, however, appeared to believe that a "prosperous" industry would soon bedrock that county.

Many Coloradans hung onto hope, encouraged as a few new discoveries were made and an older district or two revived. An example of the latter was California Gulch. Despite Hollister's ominous assessment, a few people kept placering there, while searching for the original gold deposits that they were sure must be in the nearby hills. Success rewarded their efforts with the opening of the Printer Boy Mine and its rich gold ore. A mini-rush ensued in 1868, as what was left of Oro City moved up California Gulch to near its head and settled in again. At the end of the road from Denver, it languished in the backwaters of Colorado mining.

Even with the new development, though, tiny Oro City did not promise a bright future for the territory, and neither did any of the other abandoned or nearly abandoned sites scattered about the mountains. They were too small to command attention for more than a brief moment. What kept Colorado going was the belief that not all the rich districts had been discovered nor all the rich mines opened. Somewhere in the mountains, the mother lode still beckoned prospectors, miners, and investors.

Interestingly, one myth was perpetuated even in the government documents. In a report that covered 1866, James Taylor, while admitting that Colorado mining took place "at a great cost of time, labor, capital, and skill," went on to say: "The testimony is quite general that the mine widens and grows more productive of gold at its lower stages." In his next report, Taylor was more emphatic. "A peculiarity of the Colorado gold veins is that they are invariably found richer the deeper they are sunk upon. This rule seems to be without exception and in no instance is a vein lost except by a break-off in the adjoining formation."[5] That old adage might prove true if—and it was a major *if*—the milling and smelting problems could be solved. The future of Colorado mining clearly hung in the balance, and those both within and without the territory knew it.

Even local papers joined in the criticism. Central City's *Daily Colorado Herald* (January 17, 1868) stated bluntly that 1867 "was a very bad year." It observed, "We were overwhelmed with processes of all kinds and denominations, warranted as a contemporary observed, to save 102% of the precious metal." In the October 14, 1868, issue, the editor blasted: "Colorado is a mining country cursed by processes." In one notable case, the Lyon process of lead liquation was adopted "when there was no lead in the region," making "the furnace not . . . worth the bricks used in its construction."

In 1867, the federal government started publishing reports on the western "Mineral Resources." One of the major subjects was the ore reduction crisis:

> The successful reduction of auriferous rock is a problem of the future. Quartz mining in Colorado has hitherto been unsuccessful from the failure of numerous processes and methods of desulphurization and amalgamation [report for 1866]. The process-mania, commencing in 1864 and lasting till 1867, was one of the main causes which damaged the reputations of the mines to such a degree that the country was nearly ruined by the reaction. Upon the first failure of the stamp-mills, people came to the conclusion that the ores must be roasted before the gold could be amalgamated. One invention for this purpose followed another;

> *desulphurization* became the Abracadabra of the new alchemists; and millions of dollars were wasted [in] speculations, based on the sweeping claims of perfect success put forward by deluded or deluding proprietors of patents [report for 1869].[6]

Commissioner Raymond spent a great deal of time chastising Coloradans for that "Abracadabra," their "processes, past and present." Six pages recounted the sorry details of the decade: "not practical"; "it is supposed that the process is abandoned"; "[its principal legacy] was it was neither metallurgically nor economically successful on a large scale"; "it is difficult to reconcile the history of this invention with the hypothesis of honesty on the part of the inventor"; the process resulted in "serious loss of precious metal in the slag."

Raymond took not only the mines but also the miners to task for the problems. In Raymond's opinion, "[t]he miners of Colorado were inexperienced in deep mining, and they made sad work with it at first. Even now there are few well-opened mines in the Territory." He gave advice that everyone should have heeded: "Mining is a business, full, at best, of difficulties and risks"; "it is the part of wisdom to secure at the outset as many of the chances as human foresight can perceive, or human skill control. When a man engages in this business with all the chances against him, it requires no prophet to foretell the result."

What were those chances? Bad management—"scientific men without practice and the practical men without science," "honest men without capacity," and "smart men without honesty"—made "stupendous mistakes of judgment." Greedy or naive eastern speculators, combined with "bad management, will ruin any mine." Inexperienced milling men and poor reduction methods did not help. Raymond rated bad management as the worst, though, saying that it "has done more to injure mining in Colorado than all other causes together."

Nevertheless, his report was not entirely negative. Raymond added that 1869 "marks the new era of mining in Colorado. The old spirit of idleness and speculation has passed away. The new spirit of labor and economy has sprung into power. It is beginning to be recognized that the men to develop the resources of this country are the men who live in it."[7]

Raymond's general assessment was on the mark, along with an earlier 1867 report that also held out some hope. The times were changing: Ore had been freighted across the plains and forwarded to the world's leading smelter center, Swansea in Wales, where "it might be experimented upon by the skilled experience employed there."

The man who sent the ore, and who also saved Colorado mining, was Nathaniel Hill, a respected and successful chemistry teacher at Brown University. In 1864, the thirty-two-year-old professor had been hired to travel to Colorado to look into mining properties that some Massachusetts and Rhode Island investors had purchased during the gold bubble. Concerned about why their properties were not producing the wealth they had assumed would soon be theirs, they contacted Hill for help.

The investors thought Hill, who had a reputation as a "man able to solve practical problems at little expense," would find the answers to their worries. With some reluctance about leaving his family and his college position, but with interest in what seemed an intriguing problem—refractory ores—Hill agreed to go, after faithfully promising his wife to "never let an occasion pass without writing a line at least." Off he went to Denver, by stage, in late May of 1864, and thereafter into the foothills and mountains. He kept his wife apprised of what he was doing and what he saw, in the process providing an excellent look at a depressed Colorado Territory.

> Central City is in a deep ravine. Mountains more than a mile high surround it on all sides. The gulch is so narrow that if you go into the backyards of the houses, you are in danger of falling down the chimney (June 23, 1864).
>
> The chapter of my own experience since I last wrote, is short and not particularly full of interest. I spent the time in Central City & surrounding district, in examining mines, and in becoming acquainted with the owners of mines (June 30, 1864).

He then went off with a party of men, including former governor William Gilpin, to the San Luis Valley, looking at mines along the way. Although he enjoyed the people and the opportunity to see "Mexican culture," Hill was not particularly impressed with the valley or nearby mining prospects. He then traveled to Boulder and wrote his "dear wife" that "Boulder City is the most Yankee of any settlement I have seen in Colorado." After an "exhausting climb in the mountains," he examined some ore samples. Back in Denver, Hill found himself in the midst of the Indian war that had closed the plains to travel. Reassuring his wife that all was well, he returned to Central City and eventually journeyed on to Buckskin Joe.

Hill was becoming fascinated with Colorado and its mining problems. He probably shocked his wife when he wrote, on September 4, "I bought a house and lot for $1800 in Central" and "have taken bonds of much valuable property. But they have cost me nothing and will of course be of no account unless I sell the property."

> Friday night brought us to Fair Play, a small mining town, near Buckskin Joe. A place of some note as a mining district. . . . I am now prepared to drive my business to a close. I will be prepared to leave in the third week of this month, unless I accept an offer to go back to Buckskin to examine a mine (October 3, 1864).

Returning to Central City, Hill assured his wife:

> Tomorrow I am going to a section of the country, 10 or 12 miles from here, to examine mining property. It is the last trip I will make from this section.
>
> The journey across the Plain does not trouble me hard as it is. I have "roughed it" to use a Colorado expression, so much, that I am sufficiently toughened for anything (October 16, 1864).[8]

Completely fascinated and challenged by the problems of mining and smelting, Hill resigned his professorship and returned to Central City the next year. For him, this was no "seat of the pants" effort; he studied the reduction problem from a scientific viewpoint. Although Hill rarely discussed the local mining situation with his wife, he included some detail in a letter during his 1865 visit. On June 7, he seemed in a pessimistic mood: "I felt quite blue on arriving here and finding business so dull." He elaborated:

> Mining matters are extremely dull at present. But few of the mills are running and every thing seems very quiet. To a stranger it tends to produce a want of confidence as to the future. But the fact is wages remain the same that they were when gold was worth twice as much as it is now. There is no mining country in the world that will stand 5.00 a day for labor and the product gold at par. That is particularly true in a place where but about one fourth of the gold in the ore is saved.

Hill felt that "two things are necessary to make mining here highly successful": labor at about $2.50 a day and a method of treating the ores "to save at least 50% of the gold."

Hill further noted that "The conditions which now tend to check operations does [sic] not seem to depress the price of good mining property. Such property is held as high as at anytime since my first visit." He had purchased stock, "now worth more than the estimated value." Considering all he had written, Hill may have worried that his wife was becoming concerned. He assured her: "Our mining interests are extensive and our ores are rich and with the management we have adopted, we are in my opinion, certain to realize handsome profits at an early day." All in all, "on analyzing matters I do not find any cause of discouragement."

What he had to find out in Central City was what process might work and, if it did, how to make it profitable. After trying and failing with a stamp mill, Hill returned to Providence to study ore reduction. To him, it seemed the answer lay with the "Swansea Process." The answer indeed rested there, and in February 1866 Hill left New York for Wales, where he toured the smelting plants and discussed his Colorado problems with metallurgists there.

In 1866, Hill and his partners decided to embark upon the costly and time-consuming process of shipping ore to Swansea; their gamble paid off when he returned to England and saw that the Welsh process worked. The *Rocky Mountain News* (June 28) proudly announced that the ore tested had "yielded astonishingly." Back in the States, Hill and his partners organized the Boston and Colorado Smelting Company before he returned to Gilpin County. After moving his family to Central, Hill purchased four acres near Black Hawk, a little over a mile from his home. A summer, fall, and early winter of hard work found the smelter ready to "blow in," in February 1868. The *Daily Colorado Herald* (July 10, 1868), pleased finally to have some good news, noted that "his reduction works are running night and day."

It took a while to work out all the problems, but Hill was on the right track. Initially, the matte containing gold and silver had to be sent to Swansea, but the reduction to matte could be done in his Black Hawk works. Despite complaints about a complicated price schedule and overly high charges, the inability of the process to work profitably on anything except high-grade ore, and the method being too slow and unsatisfactory, Hill was about to revitalize Colorado's mining industry almost single-handedly.

Rossiter Raymond did not criticize his efforts. In his 1870 report, he wrote about Hill's works. "The success of Professor Hill's smelting works . . . shows that the policy of consolidating metallurgical operations other than those of stamps and pans, is equally wise and necessary." The works "are owned by an eastern company and superintended by Professor N. P. Hill, formerly of Brown University, to whose excellent business management their success is largely due."[9]

Colorado mining and smelting were slowly making progress, but various other issues constantly recurred. Despite the rushes to Georgia, California, Nevada, Colorado, and places in between, the United States had no national mining law. As in 1859, each mining district had simply addressed its situation by creating district laws. This proved an easy and democratic way to solve a host of problems. Nevertheless, local rules were sometimes unclear, with vague terminology and unclear word choices. Records were kept haphazardly and sometimes even lost; the backup plan of relying on an old-timer's memory was not the best. Further, rules differed from district to district and sometimes conflicted, even though the Colorado Territorial Legislature had recognized the district laws in 1861. All this meant trouble for miners, investors, districts, and territorial mining in general and led to legal problems and expense—particularly unwelcome in the troubled Colorado mining times.

That the United States needed a national mining law was recognized in the 1850s, and a movement for it gained momentum in the 1860s, particularly after the war ended. United States Senator William Stewart, a Nevada miner and skilled mining lawyer, led the successful fight to gain a statute for lode claims in July 1866. The preamble stated:

> All minerals lands of the United States, surveyed and unsurveyed, are laid open to "all citizens of the United States, and to those who have declared their intention to become such, subject to statutory regulations," and also "to the local customs or rules of miners in the several mining districts not in conflict with the laws of the United States."

The act itself contained three significant provisions. First, it sanctioned prospecting and mining on the public domain, both retroactively and for the future. Second, it established the procedure for obtaining a United States patent for mining claims. Third, it legalized local "laws, customs, or rules of miners of the district" for such things as size and description of a claim and land used for milling purposes.[10]

Of course, the new act did not resolve all the miners' concerns, but it was a good start. One obvious omission was resolved in 1870 when an act regarding placer claims was passed, though with placer mining on a downhill trend in Colorado, the oversight had not proven burdensome. As of 1870, prospectors and miners had a uniform code to follow wherever their wanderings took them.

Rossiter Raymond pleaded strongly for the public and Washington to become involved in another issue: widespread timbering and the resultant disappearance of trees. A related problem was the destruction wrought by fire, either unintentionally started or deliberately set by persons attempting to make prospecting easier. Raymond used as an example Central City, "the oldest and most populous gold-mining center of Colorado," where the "consumption of wood for fuel is very large." As nearby wood sources disappeared, the woodmen moved further away, and the cost per cord moved upward with them, soaring from two dollars to ten dollars. The lumbermen were also robbing "Peter to pay Paul" because "they are robbing and skinning districts that may at any day require their own timber." This "reckless and disastrous" practice, Raymond stated strongly, became "one of the worst abuses attendant upon the settlement of the mining regions and other portions of the West."

"Where is all this to end?" he asked. Then he answered his own question: "And the answer of the denuded hill-sides, of the dismal wastes upon the mountain slopes, with their millions of charred trunks and ghostly whitened branches, is terribly suggestive."

He saw no easy fix. The government policies had failed, and the territorial acts were no more successful. His solution was to "put the lands in market, and sell them to settlers." "Such a plan will soon put a stop to the waste of timber. The timber once in the hands of private parties, precautions will be taken to guard it against destruction by fire; none will be cut down that is not needed; and once down, it will all be used up."[11]

Raymond spoke to the future, however. Pioneering Americans considered natural resources theirs to use as they wished. That had been their ancestors' right since settlement came to these shores; further, the development of the West benefited Washington and the entire country.

Uncle Sam's generosity overflowed during these years. Ease, comfort, and speed of transportation had been concerns of pioneers and later folks trekking west to the promised land. By the time of the Pike's Peak rush, the answer was clear, if far away: the railroad, which provided the cheapest, safest, fastest, and easiest way to travel that Americans had ever known.

Agitation had roiling for years about a transcontinental line, but the tension between North and South had prevented a final decision as to the route. During the war, with the Southerners out of the Union, a route was selected across the central Great Plains and over the mountains through the South Pass region. The task was too great for individuals or companies to subsidize, so Washington chipped in money and free land. The Central Pacific Railroad (CP) would build east from San Francisco and the Union Pacific west from Omaha.

With the war over, the pace of railroad building picked up from both ends. The Union Pacific raced across the prairie where covered wagons and stagecoaches had until recently held sway. However, horrified Coloradans were beside themselves to learn that except for "far away" Julesburg, the line would bypass their territory. Despite their promotion and protests, the railroad executives decided that no pass through the Colorado Rockies was feasible, because of the steep grade, altitude, and high cost.

The threat to Colorado's future was enormous, particularly when the UP started a rival town to the north. Cheyenne, Wyoming, challenged Denver's economic and political position as "the queen of the mountains and plains." Denverites responded quickly by organizing their own short line, the Denver Pacific, to intersect the UP east of Cheyenne so that they could completely bypass the upstart city.

They pushed their line northward while the Union Pacific raced westward, finally joining with the Central Pacific at Promontory, Utah, on May 10, 1869. The Denver Pacific had not quite reached the junction with the UP as the year ended. Nevertheless, eager Coloradans were already planning railroads that would reach into the mountains to serve the mines and mining communities. Locals up there could hardly wait. A mining boom, they believed, was near realization with the coming of a cheap, year-round, and dependable railroad.

The government's military arm also eased another problem and worry: the threat of the Plains Indians. A series of campaigns had ended with most of the tribes being placed on reservations and the opening of eastern Colorado to settlement. That provided a blessing beyond safety, because Coloradans hoped farmers and ranchers would settle there and raise crops and cattle. Homegrown crops would lower the cost of living and lessen the territory's dependence on the midwest and the east.

However, the Utes still lived beyond the "snowy range" in the valleys of the Western Slope. The various bands had already found prospectors in their lands, and some tribal leaders had an idea of what was coming. The pattern was well established: When prospectors and miners appeared, the Native Americans left, and often quickly. One had only to look to California, Idaho, Nevada, and elsewhere in the west to see this result.

In fact, the final drama was already in its first act. One of the two "new" mining districts opened in 1869 was the San Juans, right in the heart of Ute land. Charles Baker's 1861 fiasco had faded from memory, but not the legend of rich gold deposits and mines in those mountains. There had been a brief respite during the war and immediately thereafter, but in 1869 a prospecting party from Prescott, Arizona Territory, worked its way up the Dolores River looking for gold.

A second prospecting party ventured out of Central City toward northern Boulder County, where one of the prospectors had found a silver float on a mountainside. Up near the Continental Divide they discovered silver, but, being experienced miners, they said little about it to outsiders. Winter comes early at 10,000 feet, so they settled in, built a cabin, worked their claims, piled up ore, and blazed a little trail down to the wagon road between Central City and Gold Hill. They managed to keep their discovery quiet as long as the snow was heavy and the winds blew steadily. And the winds did blow; in fact, the mining camp they eventually named Caribou was known as "the place where the winds begin."

As the year and the decade ended, the outlook for Colorado's future had brightened considerably. Hill had solved many of the smelting problems and was working to resolve others. That promised to revive some of the older districts and particularly benefit Gilpin, the king of the mining counties. Several new districts were about to be added to Colorado's galaxy, and older ones had been revitalized. Coloradans knew they had, or were about to get, two of the three essentials of a new mining boom. They did not doubt that their mountains held rich ore, and the most modern transportation available was coming—the iron horse. Railroad connections had nearly been completed, and settlement promised to spread across the plains and mountains. Now all they had to do was encourage investors to come, traveling in ease and comfort previously unavailable, and invest in the wonderful mines located throughout their majestic mountains. The *Rocky Mountain News,* December 3, 1867, placed the situation in perspective when it proudly identified "Mining, our great industry, the foundation upon which rests our whole business prosperity." Those "good times a-comin" were almost at hand.

I am bound for the promised land;
Oh, who will come and go with me?

—SAMUEL STENNETT,
"ON JORDAN'S STORMY BANKS I STAND"

5

1870–1874: Bonanza! "Three Cheers and a Tiger"

The promised land! Prospectors and miners had been looking for it in the mountains of Colorado since 1859. Before the decade of the 1870s ticked into history, some at least would finally reach it. To paraphrase the words of another popular song of the post-Civil War era, it was the "Decade of the Jubilee": the 1870s, Colorado's great mining decade in which all the stars aligned. Before 1880 dawned, Colorado reigned as the premier gold and silver mining state in the United States. Not bad for a territory that ten years before had ranked behind Nevada, Montana, and California—and had seen Idaho just a nose ahead—in gold and silver production. There would never be another decade quite like this one for Colorado, nor for its mining industry.

Production statistics alone chart the astonishing change. The three old, original mining counties tell the tale even without the tally from the new districts that burst onto the scene. Between 1870 and 1880, gold and silver production in Clear Creek County jumped from $572,000 to $2.3 million; Gilpin County snuck up from $2.2 million to $2.6 million; and Boulder County, the lowest producer among the old-timers, still moved from $180,000 to $789,000. Then amazing Lake County astonished locals and the whole country alike: After having barely topped $65,000 in 1870, production there skyrocketed to $11.5 million. Only Summit County, once the placer heartland, declined in production by $75,000 to $438,000.

Total Colorado silver and gold production vaulted from $2.9 million in 1870 to $19.5 million a decade later. Clear Creek, Gilpin, and Lake Counties accounted for the lion's share of that, with Lake being the prime mover. To put these totals in perspective for the twenty-first century, gold in that decade sold for $20.67 an ounce and silver hovered between $1.35 and $1.15 per ounce, while facing a continuing downward trend.

It did not happen overnight. Like a Rocky Mountain snow slide, production picked up speed as new discoveries sparked excitements and old districts revived. Support industries matched the mining industry almost step for step, with each one enhancing the development of the other and all of Colorado gaining.

United States Mining Commissioner Rossiter Raymond correctly forecast what was going to happen and why. Railroads were coming, along with "the growth of several branches of domestic manufactures." The development of cities and agriculture taking root in her "fertile plains and parks" all boded well for mining. "It is difficult to find an instance where the two fundamental productive activities of man [mining and agriculture] are both so magnificently endowed, and so conveniently located for mutual assistance without interference."[1] In an 1870 report, Raymond noted "that the industry is now looked upon as legitimate business more than ever before. The time of wild and extravagant speculation . . . has passed away in Colorado." In this, he proved a bit optimistic, but legitimate business practices were certainly becoming more common.

The first rush of the decade was Boulder County's Caribou, whose discoverers could not long hide their good fortune when winter turned to spring. As they shipped ore out to Black Hawk, the rush started, reinvigorating the 1860s Grand Island Mining District. Claims were staked all over Caribou and the Idaho hills—and a little camp grew in the valley at their feet, following a familiar pattern. Because of the revived Nevada Comstock and its extraordinary "Big Bonanza," silver had become the darling of the mining scene—particularly as the Comstock seemed to magically produce more silver each year, yielding $61 million in 1876 alone.

Caribou even made the *New York Times*, with a multitude of headlines: "Wild Excitement Among the Adventurers of the Territory"; "The Richest

Silver Deposits on the Continent"; "A City Grown Up in a Month" (September 12, 1870). The article opened, "Ferment over this new district nothing short of wonderful. People are rushing in here from the surrounding country by the hundreds. Undoubtedly they have found one of the richest silver mining regions ever known." Perhaps this report served up a large slice of the exaggeration and overenthusiasm that were abundant in mining rush days. Nevertheless, Caribou supplied the first good mining news that Coloradans had enjoyed for quite a while.

In the years that followed, Caribou received less coverage but occasionally made brief bids for public attention: for example, when an up-to-date smelter opened at nearby Nederland, and again when the Caribou Mine was sold to Dutch investors in 1873 for $3 million, the largest Colorado sale to date. The mine's failure three years later, which stirred finger-pointing and accusations, garnered negative headlines. Then, in 1879, fire destroyed part of the community and numerous mine buildings and Caribou started its final downward spiral.

Caribou went through a complete evolutionary life cycle in one decade. It went from the wonderful days of 1870, when its "citizens believe with reason they are in one of the best mining districts of the territory," to maturity and then slowly slipped into decline. Caribou was a "Cornish" district; these skilled miners, coming over from Central City, prolonged the life of the area and probably boosted production considerably. Conservative estimates place the decade's production at slightly over $2 million, and other figures range as high as $3.5 million—but that was not Caribou's major significance. It advertised and promoted Colorado at a time when the struggling territory needed something new and promising, with no ties to old failures and scams. Like every other mining district, it provided jobs and a market for local products, encouraged further prospecting, pumped money into the Colorado general economy, and promoted (in this case) Boulder County's varied resources. Caribou lured back both investors and their money, demonstrated that Colorado still contained undiscovered mineral treasures, and focused greater attention on silver.

Through it all, Colorado's "Silver Queen," Georgetown, never yielded its crown to the upstart Caribou. Though not as newsworthy in 1870, its local production easily topped that of its Boulder County rival. At the opening of the decade, Georgetown was hailed for "more solid development" and the "energy of actual work," as it left behind the "tunnel excitement" and "desire to 'sell out'" of earlier years. By 1872, production in Clear Creek County soared over $1 million, and two years later it surged past $2 million.

In 1874, the county even became the territory's top producer—"the banner mining-county of Colorado"—before again yielding the top spot to Gilpin County, the long-time number one producer, the next year.

Like Caribou, Georgetown followed a familiar pattern. The Terrible Mine, "one of the leading producers in the county," was sold to English investors in 1870. After five years of steady production, it ran into an apex dispute and work was suspended.

Legal action, which was both costly and time-consuming, constituted one of the great blights of the mining industry. Though it gave secret joy to lawyers, the apex issue bedeviled miners throughout the west. The discoverer of a vein's apex was entitled to "extralateral rights, or the right to pursue the vein in depth beyond the side-line boundary of the originally patented surface ground, and into the land adjoining." It seemed simple. The owner of the apex had the exclusive right to mine the vein beyond his property's sidelines, but not beyond the end lines, unless he bought the neighboring claims. If the claims—and the ore—followed the orderly seams described in textbooks, all went well. They seldom did, however. Instead, the veins ran every which way and often surfaced on other claims.

With the state of mining at the time, defining a vein's apex often proved difficult and created lawsuits rather than wealthy mine owners. A noted mining engineer, James D. Hague, who worked in Colorado as early as the 1870s, wrote about this:

> How shall be determined the linear extent of the lode in depth, between end-line planes produced from non-parallel, diverging, converging, broken, crooked, or otherwise non-conformable or unstatutable surface end-lines of variable direction, sometimes becoming sidelines, while side-lines become end-lines[?]

Hague finally admitted that the law barely managed to touch "the perplexing complications of end-lines and side-lines, the puzzling identity of lodes and the doubtful place of the true top or apex." The apex, Hague went on to say, has "been the most fruitful source of conflicting interests and bitter controversy."[2] This confusing, perplexing puzzle mystified miners and everyone else, but gave lawyers, witnesses, and experts steady work and a bonanza in fees.

This legal "curse" led to numerous lawsuits in various Colorado districts, along with violence (even killings) and ruined reputations. Sometimes litigation dragged on for years before someone won or the contestants consolidated their properties. Usually, far more losers emerged than winners. The reason? According to the 1872 law, the claimant had to fix surface boundaries to include the apex of the vein; failure to do so negated the right to follow it through the side boundaries.

The Terrible Mine illustrated one possible solution. After work stopped, to avoid costly lawsuits, it was merged into a larger company. By the late 1870s, it had both company miners and lessees working the property. The Pelican/Dives, in contrast, did not fare as well in the storm that came with apex controversies. Initiated in 1873, the dispute carried into the next decade. Both sides hired toughs; armed Pelican miners seized the Dives; both western and eastern owners filed lawsuits against each other. A sad tale of bloodshed, killing, upset stockholders, and destroyed reputations was played out before the Pelican and Dives Mining Company finally emerged—as owners of a ruined mine.

Both cases hurt Georgetown's image and highlighted the complicated nature of apex litigation. The Pelican/Dives matter had been particularly costly, both in individual reputations and in money spent. The underfunded Pelican owners, for example, admitted in 1876 that they had already spent half a million dollars defending themselves. Of course, apex struggles were not the only cases. Over in Lake County, the Printer Boy Mine was jumped, and expensive, lengthy litigation ensued before the owners got it back.

The Georgetown district seemed particularly plagued by legal wrangles, however. In his 1873 report, Raymond pointed this out, along with the negative results thereof:

> The past year has witnessed an unusual amount of mining litigation, which seems to be especially the curse of this section of Colorado. The profits of some of the best mines bid fair to be absorbed in legal fees, and Clear Creek County to be bankrupted by the expenses constantly arising from "jumping," and from the bloodshed which frequently follows.[3]

Such cases made investors nervous, infuriated locals, and raised questions about other mines in the district. Unfortunately, these problems were one result of the national 1872 mining law, which was meant to clarify such issues, not to add to the confusion and encourage contentious litigation.

The rest of the 1872 mining law proved less controversial at that time. Its aim was, simply, "to offer means for a fair adjustment of thousands of claims upon all kinds of mining property, and lying between men of every class and nationality," and to develop the country's mineral resources. Of course, one had to be a citizen, or declare one's intention of becoming a citizen, to be able to stake a claim. The national law validated claim locations "by the customs, regulations, and laws in force at the date of their location" and allowed miners in a district to "make rules and regulations not in conflict with the laws of the United States" or the state or territory. In section 6, it discussed in some detail how to patent a mine. In all, the law's sixteen sections became the miner's bible.[4]

Despite the problems that Caribou and Georgetown were suffering, there were plenty of new districts opening to tempt miners and the investing public. Boulder County stepped into the spotlight with the discovery of mines that had "remarkable rich ore" at Sunshine, a "prominent little camp." This discovery, along with Caribou, "greatly stimulated mining" in older districts, such as Gold Hill and Ward, and encouraged prospecting throughout the nearby mountains.

Park County saw "rich discoveries" on Lincoln and Bross mountains, which caused an "influx of prospectors and capitalists, and called attention to the wonderful riches of the whole Mosquito Range." The "Hardscrabble district," in Fremont County's Wet Valley, centered at Rosita and Silver Cliff, also captured attention with its silver veins. That somewhat isolated district echoed a familiar refrain: It needed railroad connections, investors, and a successful local smelting works. Though the rails failed to appear in 1873, Hardscrabble passed through the "vicissitudes of early age" and awaited prosperity.

Of all the districts, however, the far away San Juan, in the southwestern part of the territory, held the most promise. This mineral "treasure house," which had started as a gold district, quickly became known for its silver mines. Despite warnings from critics who well remembered the 1861 fiasco, mining and permanent settlement had gained a toehold there by 1872. Even an unusually critical William Byers, and his *Rocky Mountain News*, which had called it a hoax or worse in 1870, praised the region by 1872. "Possibilities point to a busy and exciting season," crowed the May 18 edition. "All will be bustle, hurry, noise, excitement and confusion."

Later that year, the *News* (November 14, 1872) had an even more exciting tidbit for its readers: diamonds. Out of the past came Colorado's first governor, William Gilpin, to predict once more that diamonds would be found in Colorado and "most conclusively" in the San Juans. A correspondent wrote to the paper on December 12 that a party from the "newly discovered diamond fields in the San Juan" had arrived with a "camp kettle full of diamonds of the first water" and two "gunny sacks full of rubies and sapphires." It must have been a full moon and the lunacy season in full blossom, because to this day none of these have ever been found in the San Juans.

San Francisco's *Mining and Scientific Press* was not so sure about gold and silver, either. Its editor worried about "fearfully exaggerated" reports of ore yielding from $2,000 to $5,000 a ton, and even up to $8,000. The *MSP* finally grudgingly admitted that San Juan might draw off the drifting crowd from older districts: "It will have one good effect at least, as it will cause easier times in the old camps with high wages and plenty to do."[5] The *MSP*'s national competitor, the *Engineering & Mining Journal*, was more cautiously optimistic, but wisely warned readers of its December 2, 1873, issue, "There is a large field for prospecting, but none should attempt the trip during the winter."

Of more immediate importance was the question of what to do now that prospectors and miners were trespassing on Ute land—a development the Utes understandably did not like one bit. The Indian Bureau initially tried to deny access to the oncoming whites. The Central City *Register* blasted the attempt: "We can but regret that this decision is adverse to the development of so desirable a region." The *News*, on April 20, 1870, wholeheartedly agreed: "Our opinion exactly!" So did other Colorado newspapers.

It became a classic case of a mining rush quickly resolving the Native American/white question. As the "Ute question" came to the forefront, Colorado newspapers took the issue up and "demanded" that the region be opened, so that it could be developed to benefit the territory and the nation. Congress created a commission in 1872 to negotiate with the Utes. They failed to reach a satisfactory agreement, and the government threatened again to send in troops. A second effort, in September 1873, was more successful, and an agreement was negotiated under which the Utes ceded 4 million acres of mountainous mining land in return for $25,000 a year.

The miners received virtually all they had asked, yet trouble lay just beneath the surface. Not everyone agreed on the interpretation of the agreement. The Utes believed they had the right to hunt in the mountains, as they had for centuries, and the whites believed otherwise. Furthermore, Ute land still surrounded much of the San Juans. Neither side was truly satisfied, and the racial conflict that underlay much of the conflict still festered. The *Rocky Mountain News* warned in a bold headline: "NOTICE TO ALL PERSONS GOING TO THE SAN JUAN MINES:" do not "trespass upon any portion of the Ute Reservation by passing over or settling upon same" (April 23, 1874).

The San Juans, and the Ute land, were still rather isolated. In other areas Coloradans were welcoming some developments that would literally change their lives, and none was more popular than the arrival of the railroad.

The glorious day finally dawned when the Denver Pacific's first train at last chugged into Denver on June 22, 1870. Notably, the city's own efforts had brought this long-awaited relief. The territory's cup overflowed when a second railroad, the Kansas Pacific, hurried into Denver in August. The jubilant "Queen City" now had two outlets to the rest of the nation and the promised railroad millennium was at last at hand. Now that it had become a railroad hub, no Colorado, or other central Rocky Mountain, city would be able to challenge the "Queen" successfully.

These, however, were standard-gauge trains, running on rails 4 feet 8½ inches wide, which were not really suitable for the mountains, because of the width of the rails, the size of the cars, their problems in going up steep grades, and the cost of construction. The mountain mining communities and districts, therefore, would not receive the much-anticipated benefits of the railroad unless something could be done.

Rossiter Raymond understood the importance of railroads to mining. In his 1870 mining report, he emphatically stated, "The erection of smelting-work for the treatment of Colorado ores is a matter closely connected with the question of railway transportation." He went on to say, "The art of mining may be said to have given birth to the railway system." The mountains, however, had always presented problems because of the "unusual difficulties of grade and curve, and thus enhancing the cost of construction, while they offer in return a comparatively small amount of remunerative traffic." The answer, as Raymond saw it, was narrow gauge.[6]

That solution was nearly a reality when the two standard railroads steamed into Denver that summer. Soon a narrow-gauge railroad, only three feet wide, would be under construction. Narrow gauge obviated the problems of the standard gauge: It could climb steeper grades, go around sharper curves, and cost less to build. The tradeoff was smaller engines that could pull fewer cars, which were also smaller than the passenger and freight cars of their standard-gauge rivals.

The man of vision who put all this in motion was a Civil War veteran and experienced railroad man, William Jackson Palmer. With all the transcontinental lines building east to west, Palmer saw an opportunity. His simple plan was to construct a north-to-south railroad from Denver, eventually to reach Mexico, with feeder lines west into the mountains and east to the agricultural plains. Unlike his rivals, he did not take land from the government, but planned to make money by building communities along his route, in addition to passenger and freight traffic. Palmer selected narrow gauge because of its flexibility in mountains and canyons, and because it was familiar to his British investors.

In the fall of 1871, the Denver & Rio Grande reached Palmer's new community of Colorado Springs and then marched southward, until the crash of 1873 and subsequent depression forced construction to stop. Meanwhile, the D&RG reached the southern coal fields, and coal mining came of age, a long-needed development for Colorado and mining. Although Palmer had not as yet reached any hard-rock mining community, his railroad was poised to bring a better tomorrow.

A railroad boom was at hand. The Colorado Central out of Golden had once hoped to join the Union Pacific line, but the Denver Pacific beat it there, long before it finally arrived in 1877. Meanwhile, the Colorado Central turned its attention to a more attainable and lucrative idea: Build up along Clear Creek Canyon to the mining districts nestled in the mountains. One branch reached Black Hawk and another Georgetown in 1877, with Central City gaining a connection a year later. Appropriate speeches, happy people, and perchance a gold or silver spike to honor the occasion greeted the arrival of railheads and lines everywhere. Other rail lines were already in the planning stage, and mining people eagerly awaited their coming.

This proved to be just the start. As railroads reached out to tap the wealth of the mining districts, both prospered. Cheap transportation provided the long-awaited key to mining success and prosperity, and to the emergence of a regional smelting center. With rail access, low-grade ore could be moved economically, supplies could arrive regardless of the weather, and travel was simplified. The cost of living too seemed destined to go down.

While construction boomed on the east side of the mountains, the Western Slopers could only fretfully look forward to the "glorious day" when they could hear a train whistle. The Denver, South Park and Pacific had plans of going there, but the crash of 1873 stopped that expansion also.

Two important matters about the future of Colorado had thus been addressed: Railroad connections promised a better future, and the "Ute question" had been partially resolved. Despite these blessings, though, Colorado mining still faced some technological problems in the early 1870s.

Even with the fanfare over Nathaniel Hill's Black Hawk smelter, miners soon complained. They had long wanted a good local smelting plant and "higher prices for their ore." The cost of reduction seemed too high, the saving of gold and silver too little, and the price paid for other metals recovered too low. When railroads finally brought a large area within the economic orbit of Hill's smelter, in theory smelting charges should have fallen. Hill was forced to defend his practices on numerous occasions during the decade, despite making almost yearly upgrades to his methods and equipment. Even as he continued to improve and enlarge his Black Hawk works, he encountered his own problems, including the isolated location, the cost of fuel, and the absence of the flux needed to make his process work successfully.

Improvements in mining came in a somewhat spasmodic fashion. Steam-powered, compressed-air drills began to replace hand drilling in some mines where the familiar single- or double-jack drill had been the order of the day. There was skill and pride in being an expert single-jack miner or, with a partner, double-jacking a "round" in preparation for blasting. That professional skill would be negated by a machine. Despite some miners' refusal to work with power drills, and their complaints about danger, dust, and lost jobs, they could not stop its introduction. Rossiter Raymond, in 1870, put his finger on why: Drills "accomplish from three or four times as much as can be done by hand for the same cost, and in much less time." The tradeoff was more rock dust and an eventual health issue for many miners: "miners' con" [consumption], or silicosis.

Drills were not the only steam-powered innovation. In the larger and more productive mines, steam hoists moved miners up and down in cages; ore came up and supplies went down shafts; and, increasingly, powerful steam engines pumped water out of the mines. Safety cages, which released hooks that bit into timbers along the shaft walls if the cable broke, added a protective feature. Timbering methods also improved, but the granite of the Colorado Rockies afforded natural protection from cave-ins that many other districts lacked. Dynamite, or giant powder, eventually replaced the less-effective black powder for blasting, but the miners needed to be particularly careful if it froze.

The high elevations, isolation, and long trails to some mines led Colorado mine owners to begin using tramways, or trams, by the 1870s. Developed in California, trams proved an answer to a Colorado miner's prayer. They were like ski lifts, with buckets in place of chairs, running on cables stretched between towers. Some could be run by gravity, with loaded buckets going down bringing empty ones back up; others used steam, or later electric, power to operate.

The Georgetown area, with its high mountain mines, readily adopted trams. They eased the ever-present local concern of getting ore out, bringing supplies in, and transporting men. Though vulnerable to snow slides, costly to build, and sometimes difficult to maintain, trams were a common mine feature throughout the rest of the century. An amazed public read about one tram at the Stevens Mine, high above Silver Plume in the December 1870 issue of *Overland Monthly*:

> But what is that, a thousand feet up the cliff? A house—ye gods! A boarding-house! The glass shows us fragments of a zig-zag trail, interspersed with ladders, where the precipices are otherwise impassable. Now we see at the foot of the cliff another house, and between the two fine lines, like a spider's web, stretch through a thousand feet of air. That is the somewhat celebrated Stevens mine; the men, lumber, provisions, etc., are all carried up, and the ore is all brought down, by means of one of the ingenious wire tram-ways, becoming common in Colorado. How the mine was ever discovered I cannot say—someone must have "lit on it."

The Stevens later gained a bit of reflected fame when one of its owners, George Armstrong Custer, died at the Little Big Horn.

When it came to innovations and inventions that touched directly on their professional and personal experience, miners in the nineteenth century tended to be a conservative lot. From the start of Colorado mining, there had always been self-proclaimed mining engineers. A positive trend came with the increasing presence of university-trained mining engineers, who had learned from books instead the "school of hard knocks." Until the 1860s, Americans desiring to gain a mining education had had to travel to Europe, as no facility was available at home. In the 1860s, this began to change, with (among others) the Columbia School of Mines and the University of Michigan graduating their first classes in 1867, and the Massachusetts Institute of Technology doing so the next year. Seven other institutions joined them in the 1870s.

The Colorado School of Mines became the first state mining school established as a separate institution, with mining instruction commencing in 1873; the first student graduated nine years later. The reason for the lag in graduation reflected the education of the era. The "school seems to have been well equipped for the time, for laboratory and field work, but apparently had difficulty in securing students who were adequately prepared to undertake work of collegiate grade." Some students only wanted to take a few courses dealing with specific mining topics, and did not plan to pursue a degree program. Nevertheless, over the 1860s and 1870s a "great advance [was] made in the facilities afforded by American schools and colleges" in mining education.[7]

As mining engineer Herbert Hoover later wrote, "It was the American universities . . . [that] lifted [mining engineering] into the realm of application of science, wider learning in the humanities with the higher ethics of a profession ranking with law, medicine and the clergy." "For all the world," mining historian Clark Spence wrote, "the American West served as a kind of gigantic post graduate school of mines." Colorado stood in the forefront of this "mining revolution."[8]

The effects of the educational advances would be felt nearer the turn of the century. Great reluctance on the part of the mining community to acknowledge the skills of these college-trained mining graduates held back their full impact on the industry for a while. Trained and experienced miners were more readily accepted, and none were better than the Cornish. Thanks to centuries of honed skills, they benefited Colorado mining as few others had done. "Cousin Jacks" were skilled hard-rock miners who migrated from Cornwall to places throughout the world as the industry declined in their homeland. They had, it was said, a "nose for ore," and they brought with them invaluable experience in drilling, timbering, dewatering, hoisting, and blasting. These Cornishmen brought better equipment as well, such as the Cornish pump. Cousin Jacks also tended to recruit their countrymen to the mines in which they were working, which explains why they migrated to Central City, then on to Caribou, and then throughout Colorado.

Considering the dangers, noises, and strange occurrences that took place while working underground, it was no surprise that the Cousin Jacks carried superstitions with them. "Tommy Knockers," those invisible folk who inhabited mines, for instance, needed to be treated carefully and considerately, to ensure that they gave warnings before a mine disaster or help in finding ore bodies. Proper consideration might include leaving part of one's lunch for them; after all, if one did not, tools might mysteriously disappear, rocks might fall on miners' heads, and leads to good ore might vanish.

The Cornish might have been some of the best miners around, but they also had occasional lapses from virtue. Some owners suspected that they "high-graded" ore, or stole to supplement their income. It was said that a "mine could not be successfully worked without them or make a profit with them."

Even with the evolution of mining techniques, the advances in education, and some improvement in the smelting industry, the pace of developments within the industry did not please everyone. Rossiter Raymond was one who was not enamored of the territory's efforts:

> Whether Colorado ores are really as rebellious as in past times they have been supposed to be, or whether this Territory is supposed to be a legitimate and proper field for the trial of every new and outrageous idea that is hatched by would-be metallurgists, it is certainly true that the metallurgical industry here grows slowly, and does not show anything like the advance that is necessary to keep pace with the increasing ore-product.[9]

In the same 1870 report, Raymond expressed concern about Gilpin County having "about seventy stamp mills with more than 1,300 stamps. Probably half of this number of stamps have been in operation more or less steadily." He called for a proper system of mine ownership and management. Gilpin County was upset with such "negativism." When Raymond criticized Clear Creek County silver mines production and mills, locals rose up in their defense: "[T]he commissioner was accused of 'croaking,' and 'bearing' the silver mines of Colorado." These East-versus-West debates over mining dated from the early days and continued well into the future.

While critics were vocal about Hill and his smelter, miners and investors had even more concerns and complaints about outlying, smaller smelters that tried to resolve district problems. Generally, either the process chosen did not work well on local ores, the overenthusiastic smelter owner misjudged the amount of ore available, or the cost simply ran too high to make it profitable to work the ore that was available. Even as close as it was to Hill's smelter, Georgetown, for example, became a center of experimentation for silver smelting. Both locals and outsiders spent huge amounts of money and time in trial after trial, but each time their hopes and their investors' dreams of profits were dashed.

Not every innovation proved a blessing. For instance, "booming" gained some popularity in Summit County where placer mining was declining steadily. This environmental horror involved storing water behind a dam and then letting it flood over low-grade placer ground. The water was then channeled through a flume, which, it was hoped, would capture some gold. The procedure required few workers, worked "otherwise worthless claims," and was touted as "labor saving." Nonetheless, it left behind an ecological mess.

That was not the only environmental problem that appeared in the 1870s. Black Hawk had severe smoke problems, thanks to Hill's smelter. The Boston and Colorado Smelter roasted heaps of ore for six weeks, giving off smoke containing "considerable arsenic." Besides silencing the "song birds," the smoke potentially endangered everyone in its area.

The smoke also supplied "the cloud hitherto lacking in this morning's spotless sky." The "immense smelting works fill the atmosphere with coal dust and darkness. One mill at our right is sending forth volumes of blackness from seventeen huge smoke stacks." Black Hawk smelled much like its counterpart in Wales had when Hill visited there a decade ago. Residents were used to it, visitors were not. A correspondent to the *Engineering and Mining Journal* noted "villainous vapors, choking gases, cough-compelling odors." Other visitors recorded distinct impressions of the pungent odor that drifted down the canyons: "The road enters a gulch filled with sulphurous vapors, highly suggestive of the infernal regions." Another account described it this way: "What a glorious time pilgrims had in struggling with these sulphurous fumes, as the lumbering heavy coach dragged its way at a snail's pace up the steep grade, past huge piles of roasting pyrites!"[10]

Notwithstanding the drawbacks, not everyone thought such smoke an evil omen. The editor of the *Dolores News* (August 28, 1879) told his readers that the day that dawns on a smelter "belching forth its huge volume of smoke over the town of Rico, will be to all the inhabitants of Pioneer Mining District a harbinger of good times coming." For a new camp like Rico, a smelter represented mining coming of age and progress reaching toward prosperity.

Problems in addition to smoke quickly surfaced as well. Local streams were polluted by smelters and mines, affecting towns' water supplies. Hillsides were torn away, and trees were dying around smelters. A visitor traveling to Black Hawk and Central City, along Clear Creek, described "a stream now misnamed by reason of its dirty flood, made so by the numerous quartz mills in and about the gold region at Black Hawk and Central." Another said of the same area that the hills "seemed honey-combed or like pepper-boxes, so ragged and torn were they."[11]

Writer Helen Hunt Jackson, seeking to improve her health, moved to Colorado Springs in the early 1870s, and soon started taking tours of her new state. She eventually reached Fairplay (she spelled it "Fair Play") and was less than impressed.

> Fair Play is a mining town, one of the oldest in Colorado. It ought to be a beautiful village, lying as it does on a well-wooded slope at foot of grand mountains and on the Platte River. It is not. It is ill arranged, ill built, ill kept, dreary. Why cannot a mining town be clear, well-ordered, and homelike? I have never seen one such in Colorado or in California.

She concluded, "Surely, it would seem that men getting gold first hand from Nature might have more heart and take more time to make home pleasant and healthful than men who earn their money by the ordinary slow methods."[12] Nor was Jackson the only Easterner to be underwhelmed by the appearance of a mining community.

But what could locals do? For most, the towns and camps were merely temporary residences, and mining was why they were there, their source of income. They were using the only smelting and mining methods they knew, and they could always leave. Very few in western mining regions worried about the industry's potential impact on the land, water, air, or themselves. Some did speak out, but general environmental awareness and concern did not develop for nearly another century.

If they were put off by the reality, visitors were nevertheless often intrigued by mines' names. One trend, noticed in the 1870s, was the increasing reuse of mine names that had, one way or another, become famous for their wealth and richness of their ore. Some of the most popular ones—Comstock, Bunker Hill, Eureka, Ophir, Dolly Varden, Idaho, No Name, Seven Thirty, Homestake, and Sherman—appeared in several mining districts. In most cases, the reasons for the appellations are lost to history; perhaps miners were superstitious, were trying to fool the public or investors, or just liked the name. Still, the name of a mine often tells something about the man, or men, who originally staked the claim. Miners might choose to honor a sweetheart, hometown, home state, geographic site, historical event or person, themselves, or almost anything else; the monikers were limited only by their imaginations or the acceptable nomenclature of the era. The legendary RAM mine in Leadville dodged that latter bullet by using initials for "Ragged Ass Mine," named by the discoverer who slipped down a rocky slope and found a mineral outcropping, at the cost of ripped pants and a bruised body.

The year 1874 proved a milestone for Colorado: the first year in which the production of silver surpassed that of gold. Led by Georgetown and Caribou, and supported by a host of lesser-known districts, Colorado mining now had twin pillars to rely on. Proud local boosters optimistically forecast "every probability that the silver-yield will increase yearly." The foundation for future growth—railroads, trained professionals, skilled miners, improved smelting and milling, newly opened mining districts, new and improved mining equipment, and a growing agricultural and industrial support base for the miners and mines in the mountains—was in place. What the *Engineering and Mining Journal* (February 21, 1874) said about the San Juans really applied to the whole Colorado Territory: "It has been proved that there are some good mines here. It only remains now that capitalists should have attention drawn to this new field."

Despite the crushing impact of the economic depression that had settled over the territory and nation, sometimes optimism still ran amok in those who had contracted "mining fever." That old Nevada miner Mark Twain knew that all too well: "I am not one of those who in expressing opinions confine themselves to facts." Or, as he wrote in *Roughing It*, as he dreamed about the richness of a claim he and his partners had just discovered, "That brought the most realizing sense I had had yet that I was actually rich, beyond the shadow of doubt. For I was worth a million dollars, and did not care 'whether school kept or not!'" His contemporaries across the mountains of Colorado were entertaining similar thoughts as 1874 turned to 1875.

6

1875–1880: "All Roads Lead to Leadville"

> We knew that we were nearly "in" when corrals and drinking places and repair shops began to multiply, and rude, jocose signs appeared on doors closed to the besieging mob of strangers: "No chicken, no eggs, no keep folks—dam." We left behind us that disorganized thoroughfare called Harrison Avenue, with its blaring bands of music and ceaseless tramp of homeless feet on board sidewalks.[1]

Thus did Leadville welcome author and artist Mary Hallock Foote, who arrived in May 1879, with her mining engineer husband, Arthur, when the town was booming as none other in Colorado's short history. That summer Leadville stirred men's minds with boundless silver dreams and riches almost beyond imagination. It was the moment Colorado and Coloradans had been waiting for since the days of the '59 rush.

All that lay in the near future, however. As 1875 dawned, though, the Colorado mining situation looked more promising than it had for years. With railroads expanding into the mountains and the depression begun in 1873 ever so slowly starting to ease, the future looked bright. In his last report as mining commissioner, Rossiter Raymond noted progress on all fronts. Older districts, such as Empire and Gold Hill, which had gone through "alternate growth and retrogression," showed "signs of renewed vigor." New districts "discovered during 1873 and 1874"—San Juans, Rosita, Sunshine, and Hahns Peak—"have passed safely through the vicissitudes of early age." Several new districts had been discovered, in the San Juans, in the Gunnison Country, and in Boulder County (Magnolia). In Georgetown and Central, "the fierce litigations that have for many years retarded the development of the mines, and caused so much distress to investors, show signs of dying out."

Even the smelting industry had gained stability, although, as Raymond admitted, Colorado had been a "battle-ground for smelting companies" for the past five years. Nathaniel Hill had been successful, others were still experimenting.

Not that all the problems had been resolved—far from it. One of the problems Raymond identified was the distance that often stretched from mine to mill. Boulder County ores, for instance, had to travel down to the plains and then back into the mountains via Clear Creek to reach Hill's smelter at Black Hawk. Clear Creek County ores went fifteen or twenty miles to reach a smelter; Rosita had 190 miles "between it and reduction." San Juan "ores have to cross two snowy ranges, and spend days at least on the road before the furnaces are reached."[2]

These and other issues had long concerned Nathaniel Hill and his right-hand assistant, Richard Pearce, an experienced Cornish smelting man and metallurgist, who had come to work with the Boston and Colorado Company in 1873. He contributed several improvements to the Black Hawk smelter, but it became obvious that its mountain location was not the best. For example, coal, minerals used for flux, and all supplies not locally attainable had to be hauled up to Black Hawk, adding cost, time, and inconvenience to the operation. Additionally, the plant had reached the size limitation imposed by the narrow North Clear Creek canyon. Hill wished to expand his business and saw little hope of doing so near Black Hawk.

The answer was obvious. It would be far easier to take the ore out of the mountains to a central location with better weather, a lower cost of living, access to various mining districts, and a railroad hub. The cost of hauling ore downhill would be less than the expense of hauling everything up to Black Hawk. Hence, in January 1878, Denver leaders and Hill inspected several possible smelter locations on the city's northern outskirts. When the new Argo Smelter was blown in a year later, it was the most up-to-date plant that

Colorado, and the regional mining states, had ever seen, and the plant and processes only got better, thanks to Pearce's technological brilliance. Just as Hill, who was now a United States senator, and Pearce had forecast, the new location in Denver, Colorado's biggest and best city, with its lower expenditures for labor, supplies, and transportation, eventually allowed them to drop reduction costs.[3] Smelting had finally come of age in the Rocky Mountains.

A one-time California miner, Charles Harvey, settled in Denver just as the Argo smelter opened. Writing his family, he described a scene similar to that in Black Hawk:

> The smoke from their furnaces darkens the sky in that direction at all times, and with a wind drifts the smoke the way the wind blows so that it is a mist cloud moving along and it seems to me that about all the clouds we have here come from that smoke.[4]

People could already observe that smelter smoke killed vegetation and animals, clouded the scenery, and deteriorated human health. The toxic fumes and smoke of the Argo smelter drifted around Denver, and residents faced the same dilemma in every smelter community in the state.

At least all these smelters had material to process. The second half of the decade saw Rosita, and a few years later Silver Cliff, come into their own with both silver and gold production. Despite great expectations and the usual loud promotion, Custer County production peaked in 1880 at a little over $850,000.

Meanwhile, Black Hawk made the news for more than Hill's smelter. East and north of it in the late spring of 1878, silver was discovered in various locations. For a brief moment, it was like old times. News traveled quickly, as usual, and soon the silver strikes "attracted many prospectors and miners to their districts, resulting in new discoveries. There are now over one hundred men at work and some five hundred locations have been made." Also as usual, it did not pan out as the enthusiasts hoped.

Georgetown and Clear Creek County reached their absolute peak in silver production when the decade turned. Each mountain around Georgetown contained producing mines, but the *Georgetown Courier* was worried. Its June 16, 1877, issue complained, "it must be admitted by every candid judge that a considerable portion of our mines are not systemically worked." Add to this the badly managed ones, and one had the problems that perennially vexed many western mining districts.

Georgetown made one positive stride that same year. After years of experimental efforts, and much trial and error, a successful process to work low-grade ores profitably was developed and the local Stewart Mill opened. A visitor highly praised it in the *Rocky Mountain News* of April 17, 1877: It "is to Georgetown what Professor Hill's works are to Gilpin county, as it is the largest and by far the finest mill for the reduction of silver ores that I know of in Colorado." The correspondent also praised it for being "managed on business principles and making money for its owners." It did not quite turn out that way, but it certainly sounded good in the press, and the mill did answer Georgetown's immediate needs.

The biggest news for the years 1875 to 1879 came from over on the Western Slope, where the San Juan region continued to develop and the Gunnison Country came into its own. From both areas came complaints about the Utes, particularly since prospectors into the Gunnison district regularly trespassed on parts of the Ute reservation. The predictable result occurred, though not directly because of mining. The "Ute Problem" came to an explosive climax when the members of the White River band killed their agent, Nathan Meeker, in September 1879. After that, the San Juaners got their way. All the various bands, except the Southern Utes living far down in the Four Corners region, were driven out of Colorado within two years. Under the guidance of their agent, backed by the United States Army, the Utes left their traditional homeland for Utah by early September 1881. The mineral-rich and agriculturally promising Western Slope finally opened for settlement and exploitation by the white man.

The new Gunnison Country—Crested Butte, Ohio City, Gothic, Ruby, Gunnison—was settled not only by miners, but also by ranchers. With the news of silver strikes gaining momentum in 1878–1879, miners came in from nearby Lake City and Ouray to the high Elk Mountains and Taylor Park. The *Denver Republican* (February 28, 1880) heralded: "To the Gunnison is the all absorbing subject of thought and talk in Denver as well as on the remotest borders of the country, just now. . . . At all events the Gunnison excitement is the prevailing epidemic at present." But, just as this boom began its run, the Leadville excitement completely overshadowed it.

One exceptionally unusual evolution did take place as the region opened. Crested Butte, originally heralded for its silver and gold at the beginning of the Gunnison excitement, became even more famous for its coal deposits, which eventually benefited the entire region. Crested Butte stands alone in Colorado for changing from a precious-metal to a coal-mining district.

Meanwhile, the San Juans continued to develop slowly. Summitville became the first district to gain statewide attention. Located at the southeast corner of the region, just above the San Luis Valley, it attracted some of the big names in contemporary Colorado mining, including Thomas Bowen and Jerome Chaffee. Despite what the *Saguache Chronicle* called "almost exhaustless bodies of ore" (August 5, 1876), its gold veins proved spotty. The high altitude hampered work, the isolated location hurt, and production hit its peak at just under $300,000 as the decade turned.

Most of the rest of the San Juans looked to silver as the savior. Hinsdale County saw early excitement, with Lake City promoting itself as the "Gateway to the San Juan," because two of the major passes into the mountains, Cinnamon and Engineer, started nearby. This county, however, never equaled its neighbors during the 1870s, except in lead production. While Lake City's *Silver World* worked hard to promote local mines, the *Rocky Mountain News* proved blunter, or perhaps more honest. The June 22, 1879, paper stated that the district needed "capital for systematic development," had too little money to accomplish it, and investors seemed to have only weak faith in the prospects. Either "apathy or poverty" dogged Lake City's mines, the *News* claimed.

Beyond the mountains, isolation, poor transportation, and lack of money slowed development to a walk, at best. On the San Juan's western edge, Rico's *Dolores News* (August 21, 1879) opined that one thing, and one thing only, prevented a bonanza from gracing their district. There was, the editor claimed, "property enough to enrich and make affluent thousands of men as soon as capital is enlisted to work the mines." Between Rico and Lake City there could be heard a cry that echoed throughout Colorado (and elsewhere as well): Investors and their money were desperately needed to reach the "promised land" of fantastic prosperity.

Reality in the San Juans, though, consisted of many little camps settled in valleys, squeezed into canyons, and hanging onto mountainsides. The nearby mines looked promising, but, like the Gunnison Country, and every other Western Slope district, the San Juan denizens looked to the railroad for their mining millennium.

Earlier in the decade, Colorado had gained its long-sought and long-awaited statehood, on August 1, 1876. The constitution of the "centennial state" proved very proactive toward mining, for both present and future, by basically saying very little. There was to be a commissioner of mines. The legislature "shall provide" for escape shafts, proper ventilation of mines, and "other appliances" necessary to protect miners' health and safety, and "may" make laws regarding drainage. The science of mining and metallurgy may be taught in one of the state schools. A statement prohibiting children under twelve from working in the mines was aimed at coal, not hard-rock, mining. Most importantly, "mines and mining claims bearing gold, silver and other precious metals [except net proceeds and surface improvements thereof] shall be exempt from taxation for the period of ten years from the constitution's adoption."[5] That summarized governmental oversight of the state's major industry. The industry, and in turn the Colorado economy, benefited for years from this laissez-faire, hands-off attitude.

Colorado's gold and silver mining industry awaited something else, too: a new and major rush. By the time statehood arrived in 1876, it had been a decade since any really important district had created a boom anything like the days of '59. Furthermore, the tarnishing cloud of the "gold bubble" debacle still hung over the new state. Each new district raised hopes momentarily, caused a short-lived stir, and then petered out quickly; they all seemed to lack staying power, or else arrived too late in the decade to be properly tested. Coloradans could only look on with envy as other western mining states and territories grabbed headlines, drew investors, settlers, and visitors, and appeared set to boom past the newly minted state.

Not that they didn't continue to work, hope, and promote. One of the old districts, California Gulch, had once yielded a placer bonanza, a "long time" ago. Now it languished in borrasca (literally, "barren ground"). Migrating up to the gulch's head with the miners, who were still operating a few small mines, Oro City evolved into a "backwater" gold hard-rock camp district mostly ignored by the public and investors alike.

Among those who had left that "city" were Horace and Augusta Tabor, who had joined the rush to Buckskin Joe during its 1862 boom and operated a store and mined there until the decline accelerated. They returned to California Gulch in 1868 and joined the other 250 or so residents who tried hard to revive interest in the area.

Some Oro City folk traveled to Denver to tell the capital city newspapers about their fair land. They continually sent letters to various papers, some honest in evaluating local conditions, others eternally hopeful. Placer mining was "dull," no matter what the year, in the "pitted and scarred" gulch. However, "many lode mines are being worked," with the Printer Boy "paying handsomely" (1871). When a smelter was established at nearby Malta, an arrastra and a stamp mill fell into disuse, serving only to remind visitors of an earlier epoch.

Oro City #1 was "as dead as Sodom," and the miners who "remain have settled down for life." Wrote one correspondent, "its prosperity, like that of other mining towns about, is on the wane, and a single mine is at present its main support." In a similar vein, another writer said, "I think this county the richest in latent mineral wealth, and the most isolated and inaccessible region now known." In 1876, a local wrote in the *Engineering and Mining Journal* that probably "none in the Territory struggled against greater disadvantages." The "numberless rich discoveries remain unworked simply because there is no ore market," and, as usual, there was a "want of capital to be employed in the mines."

Despite the general inattention, a few optimists—or at least hucksters—always existed. They admitted that not much was seen about Oro City in the newspapers, notwithstanding "it is the liveliest town in the territory. Our little city 'Oro' will soon be a rival for Denver." Another hopeful writer, conceding that 1875 had been a bad year, still predicted that the "very dullest time of the year precedes the most active." He went on to forecast that there would be a "large number of visitors" coming, and it was hoped "some men with capital" who "would take it into their heads to embrace some of the many splendid opportunities for mining speculation this country abundantly offers."[6]

There was nothing unusual in all this. A handful of Colorado mining districts and camps, some even in worse shape than Oro City and its mines, could have written similar letters. However, an interesting undercurrent kept surfacing in the Oro City missives. Silver *was* being found. Henry Wood, the territorial assayer at Oro, recorded silver entries in his assay book on several occasions in 1873 and 1874. The highest had 160 ounces to the ton, but the vast majority ranged between 2 and 22 ounces. The national *Engineering and Mining Journal*, in December 1876, claimed that about $30,000 worth of silver had been mined that year. Also, mid-decade reports noted the discovery of large amounts of lead, and silver-lead ores were quite common in Colorado.[7]

All this harked back to an item that appeared in the *Rocky Mountain News* on October 10, 1860, about a "silver lode mania" in California Gulch. Locals must have known and speculated about how much silver there might be, but what could they do? They simply did not have the resources to develop those tantalizing possibilities.

Complex silver ore required a smelter. In fact, the need for a smelter to refine silver ores appeared in the *News* as early as 1872 (a silver lode with "remarkably rich" prospects spurred the idea) and again in 1873. Reaching the ore required underground mining, which took still more money, and Oro City did not currently figure in mining investors' visions of prosperity. No bonanza deposits had been found, which might have attracted both men with money and those with smelting knowledge. Transportation into the area was not particularly good, and skilled miners and mining engineers remained in short supply locally. At the end of 1876, there were also other Colorado districts that showed a great deal more promise.

Nevertheless, prospecting started early the next year, and soon reports of rich new discoveries swiftly spread throughout Colorado. On a snowy day in April, the Gallagher brothers (Charles, Patrick, and John), who had worked their Camp Bird Mine in Leadville's Stray Horse Gulch during the winter, hauled ore to the Malta smelter on sleds and received more than $3,000 for their efforts. A long letter that appeared in the *Rocky Mountain News* on April 11, 1877, undoubtedly whetted readers' interest. The Camp Bird Mine, the article stated, was "the largest and best discovery of the year, a carbonate deposit, almost unlimited, going from 60 to 3,000 ounces per ton." In neighboring Evans Gulch, a "pay streak averaging $300 per ton" had been struck.

That was enough. Investors started coming, including Governor John Routt, who purchased the Morning Star Mine and worked it when the legislature was not in session. While prospectors roamed around the surrounding hills beyond Oro City, staking claims and sinking shafts, they received two more bits of good news. A sampling works opened in April, and a month later the St. Louis Refining and Smelting Company began construction of a smelter which, by October, had opened a blast furnace. With improved weather that summer, reports of new strikes and mines flew down the mountains to the plains. Frank Fossett, in his 1879 *Colorado: Its Gold and Silver Mines,* described those days:

> In the fall of 1877 several hundred men were in the district—most of them prospecting or sinking prospecting shafts, but the amount of mining done was considerable. The ore shipments were steadily increasing in volume, and a vast amount of low grade mineral was mined for which no market was afforded. It was becoming more and more evident that the mineral wealth of California district was of immense extent. Mines were being developed on every hand.[8]

As interest grew and people began drifting in, a community, soon to be named Leadville, took root in the valley below to provide for the needs of both mines and miners. Among those arriving were the Tabors, from over in Oro City, who started the camp's second store in July. As fall became winter, excitement mounted about the rich discoveries already made, the nearby smelter that was ready to work the ore, and the camp that had been established to provide the refinements of "civilization." These Coloradans saw good times coming, and a silver cry echoed in the wind whirling around the mountains and across the foothills.

Despite the winter snow and cold, people continued to arrive. Then, as the spring of 1878 blossomed, Leadville and its mines became overnight sensations, their fame spreading far and wide. Words such as "roaring," "magic city," "fast place," "glorious camp," and "marvel" described it for visitors and readers alike. Seemingly endless numbers of people just kept coming. The population count grew from a few hundred in 1877 to an 1880 census tally of 14,820. (Some Leadville boosters stoutly maintained that the 1880 total had been undercounted by four or five thousand.) Colorado had never seen the likes of that before, nor would it again. Almost overnight, Leadville shot from camp status to Colorado's second most populous town, and some daydreamed about its becoming the state capital.

Rich strikes seemed to be occurring almost daily in that amazing, wonderful year of 1878. George Fryer staked the New Discovery on April 5, on the hill soon to be named for him. Fryer Hill proved to be the final exclamation point. As Fossett declared, "it had given Leadville its grand preeminence over any other section in the world at the present time." Lacking funds to develop the claim, Fryer sold his majority share to Jerome Chaffee for $50,000. Investors with money nearly fell over each other, as they arrived in record numbers.

It was not Fryer, however, but Leadville's first mayor, Horace Tabor, who brought fame and fortune to himself, Leadville, and Colorado and became a living legend. For years, he had been grubstaking prospectors for a share of what they discovered, without noticeable success. Finally, fortune smiled when he grubstaked George Hook and August Rische, two unpromising, inexperienced prospectors, on or about April 20. They went up to Fryer Hill, staked a claim they called the Little Pittsburg, dug down twenty-six or so feet, and hit an extremely rich vein of carbonate ore. The Little Pittsburg paid Tabor $51,000 in the first month; then $1,000,000 when he sold his share to the Little Pittsburg Consolidated Mining Company in May 1879.

Tabor's fame spread, and he, Chaffee, and Denver banker and mining investor David Moffat hosted a party of Eastern capitalists, some of whom owned stock in the company, to come and sample the wonders of Leadville and its mines that same month. The *Rocky Mountain News* was beside itself with joy: "Their high position cannot but be of advantage to our mining interest when they return and it will be a great help in placing other stocks on the New York market" (May 27, 1879).

Even the conservative *Engineering and Mining Journal* caught Leadville fever and got carried away by the grand excitement. Should the visit by these "prominent citizens," who "are very able experts and sagacious businessmen," pronounce a "favorable judgment upon the mines, the stock of this company will secure a popularity unequaled by any in this country" (May 24, 1879). They did and it did.

The Little Pittsburg had paid from the "grass roots," and Tabor soon added the Chrysolite and Matchless mines to his collection, the three richest mines on Fryer Hill. Thus was the Tabor legend born. Suddenly, almost overnight, Tabor became the symbol of Colorado mining and its bonanza possibilities, epitomizing what mining could do for the individual. Like many of his wealthy mining contemporaries, Tabor entered politics and soon was elected lieutenant governor. Unlike many, however, he poured his wealth back into a variety of ventures across Colorado, where mining profits often flowed out of the state.

Tabor's were not the only mines that captured the public imagination in those tumultuous, wonderful years of 1878, 1879, and early 1880. The Little Chief alone produced more than a million dollars' worth in 1879. The Carbonate Mine, the richest 1877 discovery, startled observers with its ore mined the next year, running 662 ounces of silver to the ton. The Glass-Pendery yielded more than 200 ounces of silver to the ton, plus iron ore. John Routt's Morning Star Mine made him a wealthy man, earning $290,000 in 1879 alone. The Robert E. Lee, discovered in 1879 and "the latest wonder of this wonderful district," yielded more than $495,000 in its first three months, a "bonanza of surpassing richness." It then produced $118,500 in twenty-four hours in January 1880. Sadly, in so doing, the owners ruined the mine by gutting the ore reserves, a fact the astonished American public learned only much later.

The frenzy was not about just silver. Gold was found in some of the mines, and the lead percentage ran high in most of the properties, which carried a value besides its market price as a flux for smelting. Though everyone had silver on the brain, 1879 Leadville included $90,000 in gold and $1.7 million in lead.

Many mines sold for six figures, and some incorporated for millions. Even though most members of the general public could not afford to own and operate a mine, they could purchase Leadville mining stock. Both the Chrysolite and the Little Pittsburg, for example, were incorporated and their stocks placed on the market. The Little Pittsburg paid seven straight monthly dividends of $100,000 in 1879, as the price of the stock jumped from $20 to $35. When the Chrysolite declared three monthly dividends of $200,000 each, or a monthly dividend of $1 per share, its stock jumped from $13 to $40. Either through dividends, or buying low and selling high, investors seemed to have found the royal road to wealth.

That was Leadville in its heyday of 1879. Leadville's mines and smelters were some of the best equipped and most modern in the country. The latest mining equipment was being used, mining engineers arrived by the score to gain Leadville experience, skilled miners found employment everywhere, and investors lined up nearly begging for the opportunity to invest in these mining stocks.

The riches were real, but investors, both foreign and American, ran a significant risk in buying western mining stocks. Among countless other possibilities, they could fall victim to watered stock, insider trading, outright dishonesty, overenthusiasm, and their own ignorance or greed. The variety of potential disasters was limited only by the investors themselves and the people they dealt with. Long odds faced them, as they sent their money to Colorado and expected that wealth would return. Nevertheless, dazzling Leadville appeared to be the perfect place to gather a fortune—to "get rich without working." Ready and willing promoters, trying to sell stock, awaited investors there to relieve them of their money before it became burdensome. Some, like the infamous George D. Roberts, misled the public time after time. Roberts parlayed that old saying about a "sucker born every minute" into a trail of broken dreams and broken mines throughout several western states.

Meanwhile, the depression days of the earlier 1870s disappeared in an instant and recollections of the gloomy 1860s seemed to be only yesterday's bad dream. Colorado had been waiting for these good times since 1859. The year 1879 surpassed even those fondly remembered days. Leadville was the talk of the nation. "Nay," some said, it was the talk of the world.

For the first time in its history, Colorado mining produced millionaires. In the earlier years, some people made money, even into the several hundreds of thousands range, but Leadville broke the barrier. Horace Tabor, David Moffat, James Grant, Jerome Chaffee, and a few others took bows on the state's financial, political, and social scene, reflecting their good fortunes. As Mark Twain witnessed with Nevada's Comstock, "There were nabobs in those days—in the 'flush times' I mean. Every rich strike in the mines created one or two." Leadville created a score or more.

Leadville was the American dream come to life. Frank Fossett caught this in his revised 1880 edition of *Colorado: Its Gold and Silver Mines*:

> Numerous instances can be related of men who came to the camp poor, and in a few months retired with a fortune. Others stay to win additional wealth, and have been almost universally successful, for money in this place, combined with experience and good judgment, has worked wonders.[9]

Mere words written on dry pages hardly do justice to the feel, the spirit, the never-ending silver atmosphere of those days. However, there was one person on the site who came as near as anyone to capturing that spirit. Few people successfully pictured Leadville at its exciting, dazzling peak when, as Mary Hallock Foote wrote, "All roads lead to Leadville." The perspicacious, genteel Foote portrayed the fevered times in insightful letters to friends back East as she tried to convey something to them that they had never experienced and probably would never completely comprehend. She captured the essence and spirit of a mining rush and a booming town at a unique moment in time, one seldom repeated on the scale of Leadville.

With silver virtually gushing from its mines, as miners dug deep into the surrounding hills, Leadville corralled the nation's attention, along with that of Foote's mining engineer husband, Arthur. Everything seemed possible and even probable, as each new day brought exciting possibilities even for this refined Victorian lady, who did not fit in with the crowds hustling along Chestnut Street and Harrison Avenue. Few of her neighbors had any misgivings about Leadville and its astounding mines. In a series of 1879 letters, Foote described what she saw about her, although her husband shielded her from some of the more sordid aspects of a town and time where the hand of society's restraints rested only lightly:

> And all roads lead to Leadville. Everybody was going there! Our fellow citizens as we saw them from the road were more picturesque than pleasing. I was absorbed by this curious exhibition of humanity all along the 70 miles long journey (May 18, 1879).

> I suppose every body here secretly wonders when they first come if there are *any* people here nice enough for them to associate with and are greatly surprised to find themselves by no means exceptional (May 28, 1879).

> I met Mr. Litchfield, one of the characteristic types of this place. He is a man of good family and education and prospects as an engineer. He gave it all up because he can't make money fast enough. He is here prospecting, living in a log cabin, has a partner, cooks his own meals, wearing a dark flannel shirt and concealing the gentleman as much as a gentleman can be concealed by such lendings[?] (May 28, 1879).

[home after a dance] Arthur looked after his pistol. People have been requested to hold up their hands even in this reasonably well governed town (May 28, 1879).

Every woman in Leadville who has a house and a husband is expected to be hospitable to the limit of her room and chairs for there are a great many young men too nice to care for the dark strata of Leadville society who have no where to spend their evenings. In return, they are the most loyal and untiring of friends and helpers (July 8, 1879).

The saddest thing about life here are the number of letters that come to us from people who made a shipwreck elsewhere and look to this place as a last chance—they are so sure when they get here life will be prosperous. It is so apparently heartless to tell them that this is no place for single women or unsuccessful men however delicately one may try to state the uncompromising truth (September 8, 1879).

The ride over the range seemed to jolt the crust off people. Affectation is almost impossible or at all events quite unnecessary where we all begin at the root of society again and build on a solid base. There is no concealing what you are in a place like this (November [undated] 1879).

We are surrounded by people who were good spirits. Young married people reunited after long separations. A few young ladies who are pleased by being in such extravagant demand. Young men making money rapidly or building houses for their wives. A great deal of nonsense and enthusiasm which could easily have been pricked but no one cared to trouble himself about his neighbor's illusions, while convinced he had none himself. Each man thought his mine was sure to pay or if he hadn't struck it he felt quite positive he would (November 1879).

The next year, in June, she wrote about the vicissitudes of mining: "Mines are invariably uncertain property. Don't you invest any of your money in mines. The only way in which I consider them safe is furnishing careers to resident employees and managers." Then she added, "this is treason from a manager's wife." Looking back over her 1879 and 1880 experiences, Foote made an observation that caught the essence of mining's impact on so many people: "The men out here seem such *boys*, to me—irrespective of age!"[10]

This Victorian gentlewoman, despite her Eastern upbringing and education, came to appreciate and be captivated by the West and its beauty and bemoaned the transformation of it. In a letter to her friend Helena, in the fall of 1880, she regretted the change about to take place:

It fills me with despair to see the rapid progress of the RR through the valley. The only lovely spot within reach—unbroken loveliness for miles—

> and now in a miraculously short time it will be spoiled with unpainted pine shanties and other accompaniments of civilization in the crude state.

Few others in Leadville would have agreed with her. They were perfectly willing to trade the beautiful valley for the benefits coming with the iron horse.

Out of her Leadville experiences, Mary Hallock Foote would carve three novels, *The Led-Horse Claim* (1882), *John Bodewin's Testimony* (1886), and *The Last Assembly Ball* (1889). Concerning the first, she wrote her brother-in-law, James D. Hague, that it has "been very difficult for me to write it and make it impersonal and at the same time characteristic." She then confessed an inner doubt, believing that the literary "field" had been "so completely and unapproachably occupied by men of much wider experience in mining camp life than I hope I shall ever have."

Be that as it may, in the opening chapter of *The Led-Horse Claim* she depicted the Leadville she had known. In a single paragraph, she captured the transformation of Leadville from mining camp to mining town:

> The discontent and the despair of older mining-camps in their decadence hastened to mingle their bitterness in the baptismal cup of the new one. It exhibited in its earliest youth every symptom of humanity in its decline. The restless elements of the Eastern cities; the disappointed, the reckless, the men with failures to wipe out, with losses to retrieve or to forget, the men of whom one knows not what to expect, were there; but as its practical needs increased and multiplied, and its ability to pay for what it required became manifest, the new settlement began to attract a safer population.[11]

Leadville's importance to Colorado and mining cannot be measured solely by silver profits. Experienced and educated mining men arrived, such as those Mary Hallock Foote knew—and they, rather than the school-of-hard-knocks, self-trained fellows, were the future of mining. The formation of the United States Geological Survey in 1879 further legitimized this new world. Clarence King, the first Director of the Survey, hoped to publish a "series of monographs which would in time include all the important mining districts of the country."

Leadville became the site of a detailed, sweeping professional examination. The Survey acted quickly: By December 1879, Samuel Emmons and a team of others were studying Leadville's geology, mines, metallurgy, and ores. Mining historian Rodman Paul called the 1881 publication of an abstract, and the 1886 full volume of *Geology and Mining Industry of Leadville*, "epoch-making."[12] This volume would be the "miner's bible" for generations to come.

Smelting also benefited from Leadville riches, as some of the best metallurgists and smelting men arrived there in 1878 and 1879. Emmons discussed this too: "In conclusion, it may be said that lead smelting, as carried on in this region, while not entirely beyond criticism, has been brought to a relatively high degree of perfection, and is extremely creditable to American metallurgists." He pointedly concluded with the observation that the success of various smelting works "has been proportional to the more thorough training in scientific metallurgy of its managers, the completeness and accuracy with which they have gauged the operations of their furnaces by tests, and the intelligence with which the results of these tests have been applied to the practical conduct of their business." Emmons hoped the lesson had been learned because it might prevent an increase in "the already very considerable number of abandoned smelters which dot our western hills and valleys."[13]

As with most Cinderella tales, there was another side to all this. A plague of insiders, who bought and sold mining stocks using private information, had the public at their mercy. The mere fact that a mine happened to be in the Leadville district often prompted promoters to inflate its stock value without any real basis for their assertions. Mining speculators, joined by promoters, ran amuck. Some were overenthusiastic, excitable promoters who saw millions in their properties just beyond the miner's pick in the drift. Investors forgot the old mining adage, "You can't tell beyond the pick at the end of the mine." Other promoters and owners were downright crooked and set out to fleece the unsuspecting investing public with fancy mine pamphlets, enticing company prospectuses, gaudy promises, and full-scale promotion. Nor could the public be held blameless. Greedy and gullible investors bought stock just because it was a Leadville mine. They read about Tabor and the other Leadville mining kings and naïvely assumed that every mine held equal promise and production capability.

The glamour of wealth also masked the inhospitable living and working conditions in a community built almost overnight. A transitory world blinked a silver beacon for all to come regardless of their experience, abilities, or health. Far too many thought it was their chance—perhaps the last—to make their fortunes. The high altitude took its toll on people's health and lives, as the number of graves in the cemetery attests. In fact, so many died that startled local boosters tried to hide the true toll in the winter of 1878–1879. The mining world was not kind to everyone. The owners and stockholders might be making fortunes, but the miner laboring deep in the mines still worked for the standard $3.00 a day, and the men at smelters for less. Many in Leadville struggled just to make a living.

Still, the excitement and glamour carried over into early 1880—and then Leadville, the stockholders, and the public received a series of severe jolts. All went well through mid-January 1880, with the Little Pittsburg already placing $850,000 into jubilant stockholders' pockets. With the stock price in the $30 range, predictions that it would hit $50 before the year was out promised even more good fortune for those lucky individuals. Then, unaccountably, the price started to dip in mid-February, going from $30 to $22, then under $20, and continuing to drop. Stock was being sold, by whom and why nobody knew. By the second week in March, investors saw it sink to a sickening low of $7.50 per share, and stockholders found themselves holding a bag of empty promises.

There were accusations of inside trading, withholding of information, corruption, and gutting of the mine for personal profit. Colorado "bears" were trying to depress the stock, shouted the eastern press. Colorado responded in kind, blaming New York capitalists, naïve investors who failed to understand the vicissitudes of mining, greedy Easterners, and wily promoters. People on both sides were charged with criminal activities, and those former prophets of profit—Chaffee, Moffat, Tabor, and others—found themselves muckraked by the press, which was having a field day with the mess. Meanwhile, the insiders swiftly departed with their profits.

Rumors, denials, accusations, and insinuations bounced from Colorado to New York and back again. Regardless, the whole story never emerged, and the parties responsible will probably never be known. There was certainly enough guilt to go around on both sides; without question, they both shared the blame for the debacle.

What did emerge out of this disaster? For Leadville, the publicity boom was busted, investors shied away, and its days of glamour vanished in the light of the scandal's unwelcome glare. Colorado suffered as well when the prestige of its most famous district waned. Just as in the shenanigan-filled, bonanza days of Nevada's Comstock, national mining suffered a large black blot on its reputation.

The dust of the aftermath of this calamity had not settled when the Chrysolite walked down the same path. March 1880 found it producing $242,000, amid the cheers of the local press, who were desperately searching for good news outside the depressing Leadville story. Stockholders pocketing their fourth dividend were pleased, too. Then the stock, which had topped $40, plummeted, in an edgy market, to the $13 to $20 range by May.

Something strange happened at this point. The Chrysolite miners marched out on strike on May 26, in a strike that soon spread to other mines, producing the first major labor dispute in Colorado mining history. Rallies, fired-up speakers, armed guards, and rumors of violence set what would become a familiar pattern in Colorado labor relations. With mines closed and threats of violence increasing, the cry went out from the owners and the sheriff to Governor Frederick Pitkin: Send in the Colorado National Guard and declare martial law! In the troops marched, and they were far from neutral in their actions. The National Guard's strikebreaking left the owners in complete control, so they reopened their mines with men who had not taken an active part in the strike.

Reported causes of this unusual event—desire for higher wages, obnoxious rules, worry about increase in hours, and the like—appeared to be a cover for something else. That "something else" apparently was the Chrysolite's financial situation. With shaky stock prices before the strike, the management passed on issuing the mine's monthly dividend, blaming the strike. Passed dividends, or moving dividend payment to a quarterly or yearly basis, rapidly undermined public confidence. However, the strike-ambushed company could not be blamed this time . . . or could it? By mid-July, the Chrysolite's price sagged from $17 to $6.75 per share. Again charges flew, including a fascinating theory that management had forced the strike to cover up the poor ore reserves and the bleak financial future that surely would attend that revelation.

The Little Pittsburg lay gutted, and the Chrysolite was being called "this notorious Colorado mine." Capital and labor had fought the opening skirmish in a conflict that would stretch well beyond the turn of the century. Mines, the district, the state, and personal reputations suffered serious, and in some cases irrevocable, damage.

Nor was that all that occurred during this tragic year. Just over Fremont Pass from Leadville lay the Ten Mile District, and its best mine of the moment, the Robinson, located near the camp of the same name. Its owner, George Robinson, entered politics, as so many Leadville mining men eventually did, and was elected lieutenant governor in 1880. After a dispute broke out with a neighboring mine, Robinson placed armed guards around his property. When he failed to identify himself during a night-time inspection, a guard shot and mortally wounded him.

Leadville's glory days of unbounded public confidence were gone, and they took with them most of the confidence in Colorado silver mining generally. The indefinable spark that separates a booming, youthful town from one slipping into middle age flickered out. Leadville's glamour and fame collapsed with its mines after only thirty months in the public eye. It was not that mining died after the embarrassments of 1880; it did not. There was only one year, during the period from 1880 through 1887, that Leadville's mines produced less than $10 million, and in that year it missed by only the barest of margins. Lead and silver continued to provide a living, but Leadville's excitement, novelty, and grandeur disappeared. The general public and investors did not care or believe anymore, and the town took on the tired, grimy appearance of something past its prime, an unattractive industrial city. Dreams of being the state capital vanished too. Promoters, investors, the drifting crowd, and newspaper reporters turned elsewhere for their next prospects and thrills.

Mark Twain and P. T. Barnum are credited, respectively, for the statements that "A mine is a hole in the ground owned by a liar," and "There's a sucker born every minute." Leadville proved that the originators of these sentiments were quite correct on far too many occasions. For a shining moment, Leadville grabbed the public's attention and ushered in the "silver eighties." Later, it heralded an even better decade in the golden 1890s. Never again would Colorado silver mining match this two-decade era, nor would the state behold a collapse like what it had just witnessed at Leadville.

7

The Silver Eighties: The Best of Times, the Worst of Times

As the tide of Colorado mining ebbed and flowed in the 1880s, many Coloradans scrambled to make a living in a world that seemed far different from that of twenty years ago. They may even have grown a little nostalgic about the "good old days." Mining had become big business, because, as one wag put it, it "took a gold mine to operate a silver mine." As the mines sank deeper, the expenses mounted, the equipment required grew more costly, the geology became increasingly intricate, and the ore generally decreased in value. The international price of silver did not help, either. It continued slipping downward and eventually broke the dollar-an-ounce barrier. (Recall that the price was for smelted and refined silver, not ore at the mine portal.) The miners could not understand why the decline was happening, unless it was an "international plot," a conspiracy to rob them of the profits from their hard work. Regardless of the cause or causes, the situation left miners squeezed in an economic vise of increased expenses and declining profits.

Placer mining had become a relic of an earlier age, although the old districts were still being reworked with limited success. Production from Park County placer deposits, for example, fluctuated from about $50,000 to slightly over $100,000 annually, as old-timers picked over districts that had long since seen better days. Boulder, Gilpin, and Clear Creek Counties each contributed small amounts as well.

At the same time, Colorado mining had its bright spots. San Juan and the Gunnison countries both came into their own, and a bright new silver district—Aspen—took the "silver queen" crown away from Leadville. Custer County's Silver Cliff, Westcliffe, and Querida enjoyed fleeting moments of fame early in the decade, when nearby silver mines kept production in the range of $500,000 to $600,000 per year; however, they collapsed by the mid-1880s. Those two old favorites, Gilpin and Clear Creek Counties, continued to produce gold worth $1 million to $2 million annually, but were hardly the newsworthy stars they had once been.

It was a mixture of the old and the new, with individual diggings steadily being replaced by company-controlled industrial mining. One could see that trend everywhere. People might yearn to return to an earlier, seemingly simpler era, but those days were gone forever.

One of the factors driving this change was the railroad, a blessing that could sometimes be a frustration as well. The century's last two decades saw Colorado railway trackage reach its all-time high. The Denver & Rio Grande (D&RG) reached the town it had created, Durango, in 1881, and a year later it extended to Silverton in the heart of the San Juans. Silverton soon became a railroad center with three smaller lines going deeper into the mountains to tap the Red Mountain, Animas Forks, and Gladstone mines.

Colorado's "Baby Railroad," as the narrow-gauge D&RG was called, arrived in Leadville in 1880, at Crested Butte and the Gunnison country the next year, and at Robinson and Red Cliff during 1881–1882. Ouray and Aspen cheered the appearance of the D&RG in 1887, Lake City did so two years later, and Creede heralded its arrival in 1892.

Competitors soon challenged the D&RG. The Colorado Midland reached Leadville in 1885, but lost the race to reach Aspen to the D&RG and had to read the *Rocky Mountain Sun*'s report of the exuberance: "At half past eight o'clock the sound of locomotive whistles was heard, and, at a signal from a rocket, additional fires were lighted on the mountains, steam whistles blew, and giant powder reverberated from hill to hill" (November 5, 1887).

The Denver, South Park, & Pacific reached Leadville initially by using D&RG tracks, but in 1884 it got there on its own. The Colorado Midland

and the Florence and Cripple Creek lines raced to Cripple Creek; the latter won the race in 1894, with the Midland arriving a year later. Meanwhile, Otto Mears's Rio Grande Southern went around the west end of the San Juans from Durango to Ridgway in 1890–1891, reaching Telluride along the way.

Aspen's *Rocky Mountain Sun* (January 2, 1886, New Year's edition) cried, "Aspen needs a railroad." That plea repeated a by-now familiar story. All districts needed a railroad, for railroads were ideally suited to mining enterprises, hauling in supplies and hauling out ore. Colorado mining districts in general were well railroaded; some were actually over-railroaded. Larger lines swallowed up smaller lines, following a trend in the rest of the industrial United States late in the century. The Union Pacific ended up controlling both the Colorado Central and the Denver and South Park. The overextended D&RG fell from William Palmer's hands to Easterners, but eventually took over both the Rio Grande Southern and Florence and Cripple Creek lines.

Mining could not prosper without railroads, but the railroads were not guaranteed profits. Unfortunately, because of high construction expenses, most lines were built on credit, financed by investors who expected profits to start rolling in once the trains reached their destination. That did not always happen. Further, the railroads were not always managed well. Far too many suffered from chancery, dishonesty, and stock manipulations, and even the most honest management often faced local, state, and national issues and requirements that could derail the company.

The railroad/mining connection was a marriage of necessity, not always love. Each needed and depended on the other, even though they often had complaints and concerns. Except for those few districts fortunate enough to have more than one railroad connection, the relationship could be contentious. Mining communities and districts did not want to be under the thumb of just one railroad, with limited train schedules and no competition for lower rates. Railroads desired continued high freight and passenger traffic, something no one could guarantee, and they pointed out that the mountain weather left them at the mercy of the elements. In the end, though, when the railroads arrived, Colorado hard-rock districts blossomed, and those left off the routes suffered both economically and promotionally.

Districts boomed in many ways once the railroads arrived. Even Leadville prospered as railroads neared it, closing the transportation gap day by day, and its production of low-grade ore benefited immensely from the availability of cheaper transportation. There was another, somewhat unforeseen, tradeoff. As one person complained, fearing the arrival of undesirables, "now we will have to put locks on the doors."

Of the three districts mentioned earlier—Aspen, Gunnison, and San Juan—the Gunnison country was the least significant and least productive. The removal of the Utes and the arrival of the D&RG happened almost simultaneously in 1881. Prospectors rushed into the area and in 1883 production topped out at $600,000—mostly in silver. The usual reports of promising discoveries, "good prospects," and "many newly discovered mines" in such districts as Tin Cup, Pitkin, Gothic, Ruby/Irwin, and Tomichi kept interest alive and the ever-hopeful on the move. However, other reports indicated that "developments so far made are not important"; "little has been done"; the district has "not enjoyed a very lively season"; the "smelter has been under very poor management" and "but little ore has been sold." "[T]he majority of the other mines are mere prospects."[1]

Several factors hindered development of the Gunnison region. The D&RG only reached Crested Butte, leaving the rest of the towns in that district out in the cold. The Denver, South Park, & Pacific tunneled under the Continental Divide, thus reaching the Pitkin/Ohio City area on its way to Gunnison. Despite this amazing engineering effort, which created the Alpine Tunnel and made the DSP&P the first railroad to burrow under the Continental Divide, the railroad arrived too late to energize local mining. The region also became infamous for its long, cold winters during which it was literally buried in snow.

Gunnison County could claim one of the most beautiful spots for mining: the Crystal River valley and the town of Crystal, which prospered briefly in the 1880s. Isolated, with a tenuous tie to the outside world along narrow canyon trails continually threatened by rock and snow slides, it never grew beyond promising. Irwin and the neighbor it swallowed, Ruby, were more productive for a couple of years, but the veins proved "mere knife blade seams, easy to lose and hard to trace out." The veins, though rich at the surface, far too often sharply declined in value with increasing depth.

Gunnison County's best years were over almost before they started. Uncertain fortunes did not attract investors, and strikes that never materialized except in some prospector's imagination or in overblown newsprint did little for its reputation.

Noted author Helen Hunt Jackson came from Colorado Springs to visit the Gunnison country in 1883, in another of her numerous forays into mining country. What impressed this reform-minded Victorian, on a warm summer day alongside O-Be-Joyful Creek, was not mining but nature's bountiful beauty. To her, the creek and its purple asters were worth far more than the coal, gold, and silver that had brought men here: "The prospectors hammering away high up above the foaming, plashing, sparkling torrent . . . do not know where it is amber and where it is white, or care for it unless they need a drink." She added that after mining was gone, "I shall still have my aster field."[2] The vast majority of Coloradans did not yet share her viewpoint. Interestingly, though, some Easterners did, creating further tension between the two groups, which were further exacerbated by some outspoken supporters of the Utes' right to their land against the wishes of Coloradans.

San Juaners were certainly not worried about such matters; they were enjoying the best decade in their history. The heart of the region—the triangle with points at Telluride, Silverton, and Ouray—produced as never before from its abundant mineral resources. If anyone ever doubted the significance of the railroad, the silver and gold production of the three counties in the 1880s dramatically showed its impact. San Juan County's production jumped from $19,000 to more than $800,000 annually, and Ouray's increased from $88,000 to nearly $900,000. San Juan topped $1 million within a year and Ouray did so a couple of years later. Even isolated San Miguel, where the Rio Grande Southern did not arrive until 1890, managed to top $900,000 that year and $1 million the next thanks to the railroad.

The Red Mountain District, nestled at the south end of Ironton Park between Silverton and Ouray, made the decade's biggest splash. Its inaccessibility, high mountain barriers, and the severity of winters had held back development until 1881–1882, and no wagon road reached it until 1883. When finally developed, the ore bodies were found to be in narrow, nearly vertical chimneys rather than in veins crisscrossing the mountains. This meant that mines such as the Yankee Girl or Guston might have high-grade ore from their "grass roots" while their unlucky neighbors encountered only barren rock. Both prospectors and investors were fooled. Seeing potential profit, Otto Mears first built a toll road, then his Silverton Railroad, in 1887–1888, to tap its trade.

Hailed as "beating anything yet seen in the state, Leadville not excepted," with "possibilities unlimited," Red Mountain gained fame as the 1880s wore on. With Yankee Girl ore running at 2,000 ounces of silver to the ton, it seemed that any nearby claim would put investors in the "lap of the Gods." With no other districts opening, the Red Mountain excitement came at the right time to draw a microscopic rush to the three little mining camps, Red Mountain, Guston, and Ironton. With a "wonderful future in store," there were many "rich" discoveries, sales of mines, and even a "fight" between Silverton and Ouray over which would be the trading center for this "booming" district. Silverton won, thanks to Otto Mears and his railroads.

Then the water problem hit: not ordinary water, but water so acidic it corroded every bit of iron it touched, eating through particularly exposed metal parts in days. Costly replacements and experiments to find a resistant material quickly cut into profits. Further dampening the enthusiasm were metallurgical problems. These ores were combined with copper that could not be worked by local smelters, so they had to be sent to Durango or Pueblo, adding further expense to the reclamation and refining process.[3]

The Red Mountain rush lasted a decade, with prospectors and investors scurrying about to find promising claims and perhaps another new district. This helped keep the entire San Juan region in the public eye.

In contrast, isolated Rico, on the western margins of the San Juans, struggled. Its ore pockets proved shallow, attempts to operate smelters failed because of lack of ore and incorrect processes, and transportation costs remained high. Though its outlook seemed gloomy, this coquette of a district teased prospectors until persistent miner David Swickhimer, who had been in the district since 1881, found the key. He believed a "second contact" of rich ore existed deeper in the earth. Although pinched financially, and dogged by mounting water problems and doubting partners, his miners (many with long overdue wages) dug his shaft downward.

Dame Fortune smiled on the effort. In October 1887, Swickhimer struck ore that ran over 500 ounces of silver to the ton. A shaft twenty feet away would have missed the ore body completely. Moving swiftly, he wisely consolidated or purchased neighboring claims, and his Enterprise Mine led Rico into the best era of its history.

Not all districts were so fortunate. Lake City and some neighboring camps had seen their best days, and already showed signs of decline. The mining industry in this district reflected many aspects—both good and bad—of the late nineteenth century.[4]

Optimism was never absent. As writer Ernest Ingersoll observed. "Everybody looks forward. Each proposes to do this and that, and to be happy 'when I sell my mine.' Perhaps this delicious uncertainty is part of the fun." However, misrepresentation and downright chicanery were also abundant. Wrote one San Juaner to the *Engineering and Mining Journal*, "We are liable to be sadly misrepresented by some of our kid glove representatives. Before we are aware of any of their designs upon eastern capitalists the damage has been done." The investors in the Golden Chicken Mine at Ophir learned this the hard way when they were plucked of their $75,000 investment.

Development money, though, was usually lacking. A letter from the little camp of Chattanooga on the south side of Red Mountain Pass, said that because the "owners are poor men," the mines with "strong veins are poorly developed." That problem could be remedied by outsiders with funds. The British arrived with pounds and crowns, got burned at Red Mountain, and then made some wise investments at nearby Telluride.

Reports of new discoveries always stirred interest. By 1889, Ouray's "gold belt had passed the skeptical point," according to David Day of Ouray's *Solid Muldoon* (September 13, 1889). As the year drew to a close, excitement grew and optimism permeated the air. *The Silverton Weekly Miner* (January 4, 1890) declared that "new strikes" were taking place in "all parts of the county." Instead of silver, prospectors now "are invariably accompanied by the hitherto unknown gold pan. The result is that many valuable gold discoveries have been made."

That news gained less attention than it should have because the big excitement at the moment was not the San Juans but Aspen, just over the mountains from Leadville. "Just over" might have been easy for a bird, but it was fiendishly difficult for humans, because those mountains are some of the highest in Colorado. Nevertheless, a few prospectors had ventured into the Roaring River Valley seeking to cash in on Aspen's silver bonanza.

Reports from the district seemed encouraging in 1880. Prospects were being developed, two little camps had started—Ashcroft and Aspen—and ore was being piled on dumps "there to remain until the smelter starts up next summer." By 1881, three newspapers were already promoting Aspen's mineral and mining wonders and telling the world it was the place to invest, as well as defending it against jealous rivals.

Every mining district needed loyal newspaper editors. For example, the *Rocky Mountain Sun* (July 22, 1882) gave high praise to the Smuggler Mine, which had opened with new machinery and a large force of men that past spring. "Work steadily has progressed day and night since," with "determination to prove up the property at every point." "[G]reat enthusiasm prevails among both the miners and owners of this property." In this case, the editor was not just puffing a pie-in-the-sky prospect.

Almost anything that happened in those early days elicited favorable comments. Back on November 5, 1881, the *Sun* had encouragingly reported that a smelter was being built. That was excellent news, but Aspen "needs more. In this age of improvement and progress new inventions are coming forward each day and it becomes us to see that Aspen has the advantage of all of them." Even so, the government mining report for 1882 noted that "the mines of Aspen, taken as a whole, show but inconsiderable development." The "very high-grade ore" was still being "laid on the various dumps to await the necessary machinery, as there are no smelters in operation up to this time."

Aspen's neighbor and early rival Ashcroft, the "mining wonder of the west," gained publicity as well, particularly when Colorado mining legend Horace Tabor purchased the Tam O'Shanter Mine and nearby other properties. It did not hurt the district when Tabor replied, when asked about this property, "I'm afraid to go and look at this big vein. Everybody who's been up there has come back crazy. I think it's safer to stay here, if I want to keep my head" (*Denver Tribune*, September 6, 1881).

Tabor, who had also invested in the San Juans, inspired Aspen locals to hope that "where the big fish go, the little fish will follow." What followed, however, was a costly lawsuit over the title and sale of the mine, which Tabor eventually won. Tabor found out, here and elsewhere, that being rich and famous attracted lawsuits by the score. A road and smelter were constructed, but Tabor's mine was a chronic nonproducer. Indeed, neither the Tam O'Shanter nor the other mines near Ashcroft proved to be even close to "mining wonders of the West."

That title better fit neighboring Aspen, Colorado's new silver queen. After typically slow production during its first couple of years, Pitkin County topped $1 million in silver production in 1885. That figure dropped until the railroad arrived, then hit full stride and produced more than $2 million per year for twenty years through 1907. In two of those years, 1891 and 1892, it topped $7 million.

A major reason for the slump in the mid–1880s was that old bugaboo, apex litigation. That type of litigation seemed to be a Colorado epidemic, particularly in potentially prosperous districts, where owning an acre at the right spot could make an owner a millionaire. With such high stakes, some of the best lawyers available took on apex litigation; the trials garnered statewide and sometimes national attention and were repeatedly covered in mining journals.

The Aspen fight, which started in 1884, went on for four years, but even before that the district had seen its share of lawsuits over claim boundaries and other issues. This one, however, started with the Aspen Mine, the camp's first great bonanza and the principal factor in the upsurge of production in 1885, when $600,000 came from the mine. David Hyman, owner of the next-door Durant Mine, thought that perhaps the wealth had come from his property, and so the lawsuit commenced. Whoever owned the apex would be king of the hill, and on the answer rode potential millions of dollars. For the moment, other cases were suspended, waiting to see how that one came out.

Aspen suddenly found itself in the midst of a crisis that could ruin it. Were the town and district being held back by selfish owners? The *Aspen Daily Times* (February, 24, 1885), worried that development of the district would be hindered, reported that "[t]he people are wild with excitement and indignation." Both sides lined up the best lawyers available, mining engineers to provide "expert" testimony, and a few locals to provide tidbits, opinions, and gossip. Both sides also collected maps, made models, and scrounged every conceivable bit of evidence that might prove or disprove a point. Detectives were hired to check on prospective jurors. The ultimate prize made all this expense worthwhile.

The state court trial did not commence until November. Finally, in December 1885, the jury rendered a decision in favor of Hyman, but the case was then appealed to the federal court in Denver. Eventually, with both sides having spent hundreds of thousands of dollars, common sense prevailed and they reached a compromise. It was worth the effort; the Aspen Mine proved even richer than hoped.[5]

Not only had the case been tremendously expensive, it had also created a gulf between the two parties and their friends, and the mines generally sat idle awaiting the outcome. Reputations were made and lost. Said Hyman, who surely ought to have known, "Having a knowledge of law and litigation I can say truthfully that no litigation equals a mining litigation in its intensity and bitterness." Other mining districts throughout the West watched the case with interest, knowing that they too might be caught up in their own apex fights, which, as the *Engineering and Mining Journal* told its readers, were almost always "complicated case[s]."

With the Aspen disputes resolved, mining could finally resume. The *Rocky Mountain Sun* welcomed visitors to "one of the most picturesque and peaceful valleys in the state." Without question, the "prospects [were] very bright" for both business and mining. The paper also proclaimed that "farmers were coming," but it did not go into detail about what they might raise in the high valley with its short growing season.

Mining there exceeded expectations in the years that followed. The *Engineering and Mining Journal* (March 9, 1889) told of a smelter finally testing "favorably" with local ores, adding that "soon mine owners in this valley will be able to get ores smelted without freighting expense." With a successful smelter, "thousands of dollars would be added to the wealth of the community every month." Meanwhile, the Mollie Gibson shipped more than four tons to Denver smelters, which sample assays "indicate[d] a run of 20,000 ounces to the ton."

The story proclaimed that "a genuine mining fever seems to be taking possession of this district . . . the excitement in Aspen is said to be akin to that of California in El Dorado days." Perhaps, but another trend was also obvious: The mines were coming under the control of fewer people, and Aspen was becoming an industrial mining community just like Leadville.

That was one reason the Mollie Gibson was so important to the district's image. Not only did it actually live up to its billing, but more importantly, its owner, Henry Gillespie, was that favorite of Colorado mining legend, the "rag-to-riches" story. With the best of his generation, he knew how to promote his property and raise much-needed money from investors. First, he put on exhibit a 300-pound lump of ore assaying at a reported 10,000 ounces to the ton. Then, in May 1889, he displayed a 1,000-pound monster carrying a 6,000-ounce label. The *Aspen Daily Times* (February 24, 1889) could not contain itself, crying "WEALTH! WEALTH!" With the public's interest whetted by fall, Gillespie and the other owners of the mine organized the Mollie Gibson Mining and Milling Company with stocks for the public to buy, but it did not work out. The stock value went down, not up, and Gillespie saved himself and the mine only by merging and placing the company under different control. The Mollie Gibson was not finished, however.

The arrival of the railroad had been the key to Aspen's emergence as the new "Silver Queen"—once the lawsuits settled down. The railroad allowed mines with low-grade ore to be worked, with an increasingly large output of lead, which allowed some silver mines to continue working. By the turn of the century, Aspen's lead production topped $1 million, nearly surpassing the total from silver.

While the Mollie Gibson, the Aspen, and the other mines promoted Aspen, silver mining, and Colorado, they briefly masked a crucial trend: the continuing decline of the price of silver. As mentioned, the price paid for pure silver had dipped from $1.35 to less than $1.00 per ounce. Initially, this hit the older districts harder than Aspen, because their bonanza days had long vanished, and as they went deeper into lower-grade ore, their expenses went up.

Not only did expenses increase, but profits and dividends declined or simply disappeared, discouraging the desperately needed investors. Why should they invest in mining when better returns could be had elsewhere in the booming American economy? Other than locally, Colorado mining was no longer newsworthy. It became more of a tourist attraction than a miner's and investor's paradise.

Caught in this economic vise, Colorado silver miners decided that the villain was actually national monetary policy. The tale went back to 1873, a fateful year in which a dastardly deed had been done.

In its simplest terms (in which the silverites were prone to cast it), the issue pivoted on the use of silver in coinage. Unlike gold, the market price of silver had fluctuated for decades. That was not the case with Uncle Sam, however. The federal government maintained a ratio of 16 to 1; that is, sixteen ounces of silver equaled one ounce of gold or $1.25 per ounce for silver with gold at $20. Washington believed it had to preserve that ratio to maintain the value of its silver coins. On the international market, however, that policy undervalued silver.

The predictable result ensued. Silver coins almost disappeared, melted down because the value of the metal was higher than the coin's face value. Congress responded in 1853 by ending the coinage of silver except for the dollar. Few Americans noticed or cared, but this put the country on a de facto gold standard instead of a two-metal basis. Then, during the Civil War, the government printed paper money, the first "greenbacks." The public had to have faith in the issuing government because the bills had no intrinsic value. In peacetime, they might cause inflation, whereas gold, with its steady value, would not.

After the war, businessmen, eastern bankers, and eastern investors wanted to retire the paper money and return to a "sound money" gold basis. They did not want to accept inflated currency for their products or to be repaid with cheaper paper money what they had lent in hard currency. Against them stood Westerners and Southerners, often debtors, who wanted to pay back the gold they had borrowed with inflated-value paper money. Their creditors charged that this would "dishonorably" shift some of the debtors' burden to the creditors, who were being paid back with "cheaper" money.

At the start of the 1870s, silver miners who sold on the international market and not to Washington had no such concerns. Silver coins had all but disappeared, replaced by paper money or gold coins. In 1873, Congress made the obvious response and dropped the silver dollar. Few predicted, and even fewer understood, what that action would entail.

That apparently innocuous move became the infamous "crime of '73," the silver miner's worst nightmare. Thanks particularly to the Comstock's "Big Bonanza," and to a lesser degree to success in Colorado, U.S. silver production shot to record highs. At the same time, several countries switched to the gold standard and stopped buying silver. Negligible industrial and commercial needs could not pick up the slack, and the price of silver dropped below $1.20 an ounce in the year of Colorado's statehood.

Not worried initially, the unsuspecting miners turned to Washington, only to find that Washington was no longer buying at $1.20 an ounce—or at any price. The situation then became the "crime." The silverite "cause" was born of desperation and dismay. The miners wanted two things: resumption of silver coinage and a government-guaranteed silver price of $1.25 per ounce. It was an inflationary idea that struck horror into eastern bankers, financiers, and conservatives. Nevertheless, western congressmen, joined by others from farm states where inflation might help hard-pressed farmers, managed to pass the Bland-Allison Act and to re-pass it over a veto by conservative President Rutherford Hayes in February 1878.

It seemed like a victory at first. The Secretary of the Treasury was ordered to purchase from $2 million to $4 million in silver, at "market price," each month. Miners now had silver dollars, but not at the 16-to-1 silver-to-gold price they had sought. Further, conservative Treasury Secretaries saw to it that the government purchased the lower amount, so that these dollars would "be as good as gold." Treasury secretaries of both parties were ready at all times to redeem silver dollars, whatever their "intrinsic value," in gold.

Except for the new purchases of silver, the silverites' campaign had been completely defeated. The issue became contentious in western silver-mining states, including in fervently committed Colorado. Where a candidate stood regarding the cause meant more on Election Day than anything else.

There also had to be villains in this "great tragedy." They were easy to find: eastern bankers, foreign bankers, English bankers, eastern investors, their congressional spokespersons, a "Jewish" banker conspiracy, and anyone who wanted to be repaid in "God's money." Apparently God favored gold, although the silverites repeatedly pointed out that the Bible mentioned both metals. Second Chronicles 9 recounts how "all the kings of Arabia and the governors of the land brought gold and silver to Solomon," and Jeremiah 10:9 tells that "beaten silver is brought from Tarshish, and gold from Uphaz." To the silverites, this proved that God intended both for man's use as money. Easterners could not believe their western neighbors' fanaticism. To them it all seemed emotional, illogical, based on fantasy, desperation, debtors searching for an out, miners grasping at silver straws, eager politicians looking for a cause—and it was all those things, and more.

Throughout the decade, Colorado's representatives and senators led the fight to save silver, no one more fiercely than Senator Henry M. Teller, who became Colorado's silver spokesman. He proposed this resolution on March 21, 1882:

> That the experience of mankind has demonstrated the necessity of the use of both gold and silver as circulating medium. That the destruction of the money faculty of silver is in the interest of a few only and not calculated to benefit the great mass of mankind. . . . Resolved: That it is the duty of the United States government to provide a coinage capacity equal to the annual domestic production of gold and silver.

A few excerpts from his speeches further illustrate the tenor of the times.

> [January 19, 1886, speech] I do not speak in the interest of any special classes of laborers or producers but in the interest of all who toil, labor, and produce; who create by their labor all the wealth of the world; the people who do not stand in high places, whose voices may not be heard in bankers' conventions.
>
> While not possessing all the culture of the age, they exemplify the many sterling virtues of the race and do much toward keeping alive sentiments of honesty, justice, and truth—as mighty in numbers as modest in aspirations.
>
> What I do demand is that it [silver] shall have an equal chance with gold as a money metal.[6]

Horace Tabor joined the fray as a very interested party. "Why should America, the greatest producer of silver, allow any foreign power to set the price of silver that they are to purchase from us? We have toadied to the English too long." In a speech on January 30, 1890, he sounded a clarion call:

> Let silver drop to 75 cents per ounce and there would not be a silver mine worked in America, and all our western cities would be paralyzed; our railroads would cease to pay. The laborers who are directly and indirectly interested in mining would be without employment and they would have to get away as best they could and go to the over-crowded cities of the East, leaving their homes (for most of them have acquired good and comfortable homes) for the occupation of bats. I venture to predict that with silver at $1.29, our farmers will never sell their products so cheap again. We demand the money of the constitution, the free coinage of both gold and silver.[7]

Both men foreshadowed ideas that reappeared in William Jennings Bryan's more famous "Cross of Gold" speech.

Colorado newspapers joined in, often with more emotion than logic. Even before the 1880s, they had issued blasts. The *Rocky Mountain News*

(September 20, 1876) blamed the "money kings of Europe" for the "demonetization of silver." The *Silverton Democrat* (February 13, 1886) protested that the "money shortage" meant "dull times and low wages."

"Every citizen of Colorado ought to feel a lively interest in maintaining this prosperity, and in encouraging the production of silver. How a citizen of Colorado can entertain other views and still believe himself a friend in the interests of the State, we cannot readily understand," bluntly stated Georgetown's *Colorado Miner* (October 30, 1880). A year and a half later, on April 27, the same paper called "demonetization a national calamity." In Ouray, editor Dave Day, through his *Solid Muldoon*, repeatedly and vigorously claimed that Ouray, the San Juans, local mines, and just about everybody else were being hurt by "greedy" Easterners and bankers.

The eastern press fired back at these western radicals, who threatened the very fabric of the country, its monetary system, and its prosperity. The *Engineering and Mining Journal* voiced its opinion, albeit with less emotion, in its February 7, 1885, issue. This article pointed out that the silver uproar was "seriously alarming the solid business men of the country," adding "that the United States could not maintain a silver price by itself," unless "some agreement [could] be made with other civilized governments for the establishment of a double standard." To do otherwise would "be the height of folly" and would "expose ourselves to the disasters that the continued expansions of our silver coinage expose us to."

The editor did make an attempt at placating the silverites, saying it "is our desire naturally to promote the best interests of our silver producers." It did no good, though: The lines had been drawn and the country was split. Republicans and Democrats, farmers and miners, rural folk and city slickers all joined the fight over the "cause." The few "gold bugs" in Colorado, if not nearly eternally damned, were unquestionably shunned.

As Americans are wont to do, the silverites organized. Their National Silver Convention held meetings, published papers and reports, sponsored speakers, and generally helped keep the issue at the forefront. The cause took on new meaning almost every day; they were fighting for their very livelihood, for the America of their parents and their youth. Against them were arrayed the hordes of urban, industrial America—a world they did not completely understand or comprehend and certainly did not yet share.

Their great spokesman, William Jennings Bryan, summarized the silverites' feelings when he emerged as the thundering voice of a fading way of American life. In humid, sweltering Chicago on July 9, 1896, he reached an apex with his "Cross of Gold" speech to the delegates to the Democratic

National Convention. Few mining men were in attendance that day; nonetheless, Bryan still spoke for them: "The humblest citizen in all the land, when clad in the armor of a righteous cause, is stronger than all the hosts of error." Enthusiastic agreement with their cause was just, even "holy."

Turning to the "gold" Democrats in the convention hall, Bryan issued a challenge. "When you come before us and tell us that we are about to disturb your business interests, we reply that you have disturbed our business interests by your course. We say to you that you have made the definition of a business man too limited in its application."

A few moments later, Bryan turned specifically to grievances of Colorado and all western mining states: "The miners who go down a thousand feet into the earth, or climb two thousand feet upon the cliffs, and bring forth from their hiding the precious metals to be poured into the channels of trade, are as much business men as the few financial magnates who, in a back room, corner the money of the world."[8]

Those comments revealed the core of the issue. It was "us" versus "them," the rich and powerful greedily taking away the livelihood of the "humblest citizen." As the 1890s opened, Colorado miners and mining girded for the fight of their lives.

All join 'round and sweetly you must sing,
And when the verse am through,
In the chorus all join in,
There'll be a hot time in the old town tonight.
— JOE HAYDEN, "THERE'LL BE A HOT TIME"

"There'll Be a Hot Time"

From the horse to the horseless carriage. From the candle to the electric light. From the train to the plane. From the stage to the motion picture. From the telegraph to the telephone. From rural farm to teeming city. From a wilderness territory to a state. A teenager in the Pike's Peak rush would have seen it all by the time he or she reached a seventieth birthday. By then, as World War I neared, several mining booms and busts had occurred, once-bustling camps had become ghost towns, and that teenager's generation had been hailed as pioneers.

It had been a "hot time" in the "old" towns of the Rocky Mountain country. *Marvelous, amazing, shocking, unbelievable*: these were only a few of the words used to describe those years. Americans watched their pace of life pick up and their world simultaneously shrink and broaden. Change occurred everywhere in the mining West, but particularly in the towns.

Seduced by both its beauty and reality, artist/writer Mary Hallock Foote painted a word portrait of her Leadville home.

> The Leadville scene, when it wasn't snowing or sleeting or preparing to do both, was dominated by a sky of so dark and pure and haughty a blue that "firmament" was the only name for it.
>
> [Nothing marred the beauty.] Indiscriminate houses dropped here or there did not count, nor smoking chimneys of smelters, nor brass bands every evening, if you lived far enough from the streets, they paraded to get the sound softened without sounds of the human surf beating on the flanks of those gulches.
>
> Leadville was one of the wildest and also one of the most sophisticated of mining camps.[1]

That was Leadville in 1879. Americans were intrigued, repelled, amazed, and stunned by what they read about the goings-on in Colorado's urban mining communities. There was change everywhere, however, not just out in the Rocky Mountains, as the United States raced into the Industrial Age.

Perhaps because they lived in a world of constant and accelerating change, many Americans became fascinated by the relics of an earlier age that littered the Colorado mountains and valleys. In many places, boom and bust cycles had left only a few locals still hanging on to the hopes and dreams of yesterday. As these old-timers looked at the sagging buildings, shuttered windows, rotting board sidewalks, and rats scurrying about where people recently had walked, talked, and danced, they could only remember with melancholy what had once been. Visitors, though, could imagine and speculate before hurrying through and leaving the fading past behind.

These camps and towns had once been a fast-moving, exuberant urban world—an urban West, not the more typical slow-moving, slow-settling rural frontier of Midwestern fact and legend. In reality, the mining West had never been a frontier; the initial settlements had brought near-instant urbanization. This phenomenal urbanization/mining conjunction could plainly be seen in towns such as Aspen, Telluride, Leadville, Georgetown, and Central City. The more numerous smaller camps, such as Animas Forks, Caribou, Buckskin Joe, and Schofield, dotting the landscape almost all had dreamed of becoming towns. Many had tried to jump-start the process and call attention to themselves by adding "City" to their names. Despite their best efforts, however, the vast majority did not evolve much beyond the first months after their birth.

What was the difference between a *town* and a *camp*? Towns had a larger population, a greater variety of businesses, permanent architecture beyond log cabins and one or two frame buildings, richer mines nearby, and a spirit and confidence that did not exist in a camp, no matter what camp residents might claim. Camps had a general store or two, generally rougher architecture, less prosperous mines, and a smaller population. Many of the camps teetered on the brink of becoming "ghost towns" as early as the late 1870s, and by the turn of the century most of them had become relics of the past rather than promising hopes for the future.

These camps and towns, however, framed the picture of mining for the tourist public, and often the reading public as well. Dark mines deep in the earth might be fun to visit once or twice, especially if one owned stock, but in boom times the mining communities offered excitement, tolerance, and a flair rarely seen in Victorian urban America. They had one foot in the America of an earlier age and the other in the erupting industrial America of the late nineteenth century. They represented some of the best and some of the worst aspects of a country in the throes of transition.

Helen Hunt Jackson tried to give her eastern readers an impression of a mining camp:

> It is hard to keep separate the fantastic and the sad, in one's impressions: hard to decide which has more pathos, the camp deserted or the camp newly begun, the picture of disappointment over and past or that of enthusiastic hopes, nine out of ten [of] which are doomed to die. I have sometimes thought that the newest, *livest*, most sanguine camps were the saddest sights of all. . . . The expression of a fresh mining camp, at the height of its "boom" is something which must be seen to be comprehended.[2]

Leadville in particular, at its amazing peak of wealth and electrifying pace of growth, fascinated Americans country-wide. No other Colorado mining town either of its era or later—except perhaps Cripple Creek—ever matched its mystique, captured in an 1879 pamphlet, *Leadville and its Silver Mines*:

> LEADVILLE! The very name has, within the last few months, become synonymous with sudden wealth and wild dreams. Tens of thousands of persons between the Alleghenies and the Pacific coast, and many no doubt on the Atlantic coast, are turning their feverish eyes toward this place, and saying, "Were we but in Leadville, we might strike a fortune to repair the disappointment of our past lives."
>
> The town has now—no one can tell what it will have next month, people are pouring in so fast—six well defined streets. The population to-day is about six thousand, moderately estimated; next September it will probably be ten or twelve thousand. It is the highest city in the United States.[3]

People elsewhere in the country could hardly believe or grasp what was happening in the "Cloud City," where fortunes could be made almost overnight on Chestnut or Harrison Streets and on Fryer Hill.

Coloradans of those peak mining years had no idea they were living in such romantic, enthralling, legendary, and captivating times. Many were no doubt astonished to have eastern journalists and newspapermen portray them in such terms. They were equally surprised when tourists arrived to see the "elephant" and savor the times. They were, after all, very utilitarian communities, established for one reason and one reason only: mining. Beyond that, anything else that transpired was a bonus.

Small communities were found everywhere in the mining regions—perched on mountainsides, buried deep in canyons, huddled in valleys, and nestled along creeks and rivers. Visitors such as Isabella Bird were either fascinated or repelled by what they saw. She indicated her feelings about American towns when describing her arrival in Georgetown:

> The area on which it is possible to build is so circumscribed and steep and the unpainted gable-ended houses are so perched here and there, and the water rushes so impetuously among them, that it reminded me slightly of a Swiss town. It is the only town I have seen in America to which the epithet picturesque could be applied.[4]

Quickly built in log and frame, camps appeared on the heels of each rush, large or small. If all went well and the mines developed, a camp might survive a season or two, or even a decade or longer, and perchance even grow into a town. Residents lobbied hard for their towns to become a county seat, a status that promised a stability not otherwise available. Many survive today for that very reason.

The camps and towns were there to serve the mines and miners in their district. The towns usually had satellite camps near them, which town residents considered part of their business sphere. When the inevitable decline came, camp residents drifted away or into the town, their former homes abandoned to deterioration. Towns were very protective and were jealous of interloper rival towns trying to take camp business away from them. Such maneuvers led to more than a few newspaper tempests.

A town usually had one or more newspapers, whereas a camp had only one, if any. Newspapers were critical for a community: They promoted, defended, attacked, reformed, motivated, aggravated, reported, and dissected events and activities. Every camp and town wanted one; without a newspaper to promote local highlights, miners were isolated and adrift in an uncaring world, lacking the ability to capture public notice, attract capital, or intrigue investors. Most papers had decided political slants, opinions, and editorial attitudes. They all debuted with high hopes and fair promises:

> We shall supply the need of the San Juan Country for a paper. We shall preach no sermons. We shall not dabble in politics. There will be nothing pressing upon our time or space to prevent giving full and complete reports from this and the adjacent mining districts (*Silver World*, Lake City, June 19, 1875).
>
> With our politest bow we greet you, dear reader, hoping for a long, agreeable and profitable acquaintance. We propose.
>
> 1. To aid in developing the rich mineral resources of these mountains
> 2. To Furnish the dwellers of these mountains the very latest news from the east.
> 3. In politics we are Independent Republican
>
> (*Tri-Weekly Miners' Register*, Central City, July 28, 1862).
>
> We have opinions of our own and want to express them. We have matters of interest to tell and want to make them known. We have a work to do and must be about it. Our purpose is to work for the *best interests* of all our people, without any limitation or reservation whatsoever. Furthermore, we have come to stay (*Crested Butte Republican*, October 5, 1881).

Eventually, most of these papers suspended publication, usually with a valedictory editorial in the last issue. The *Caribou Post* (August 17, 1872) explained to its readers, "We started it [*Post*] fifteen months ago, largely in anticipation of the future." Development did not prove as "rapid as we had reason to hope and expect, and we have consequently been unable to make it a present financial success." Thus the paper bid readers farewell: "That we are compelled to suspend at the present time, is not the fault of anybody, but rather the misfortune."

Tarryall's *Miners' Record* folded for similar reasons. In its last issue (September 14, 1862), it explained simply and honestly that "receipts do not pay current expenses." The editor complained that people had "manifested but little disposition to advertise" and job work had been "scarce." These two factors, the editor pointed out, "keep the newspaper office alive." He held out hope for the future, though, as he bid his readers "adieu for a time."

Mirroring the unbounded optimism of their readers, the proliferation of newspapers often outpaced the number of subscribers and advertisers needed for success. Some of the larger towns had too many papers for the available readership. Politically, papers usually ardently championed a single side, thereby potentially alienating a goodly proportion of the possible readership and putting themselves on the thin edge of profitability. The newspaper business in the Colorado mining world proved a gamble at best. Editors had to wear many hats and diligently search out the news, promote their communities, and sometimes even defend themselves. Nevertheless, they and their papers generally served the towns and districts well, and today they provide an invaluable window on the past.

The other group that had great stakes in the future of communities was the businessmen and women. They invested more in the present and future than did the transitory community members; it thus behooved them to work for permanency. Their numbers could be impressive. During Leadville's plush days of prosperity, in June of 1879, the town was home to 25 clothing stores, 17 barber shops, 44 hotels and lodging houses, 51 groceries, 20 meat markets, 31 restaurants, and 120 saloons, among other businesses.

No other Colorado mining town could claim such numbers. More typical was Telluride in 1884, with its ten saloons, two hotels, five mercantile stores, and two restaurants. The size of and variety in the business district constituted a major difference between towns and camps. For example, Animas Forks, over the mountains from Telluride, had one hotel, two general merchants, and one restaurant in 1882. Breckenridge, still in its camp stage in 1861, boasted several hotels, stores, and meat markets and the only post office "west of the range."

The ubiquitous general store, which offered a little of everything but not much selection, appeared everywhere. If the owner also gained the local post office, she or he had a decided advantage over competitors for miles around. Most business districts also had a livery stable or two, sometimes even more. That establishment, along with the saloon, was the "man's home away from home," good for a discussion, a game of checkers, and maybe a peek at the "racy" *Police Gazette*, with its buxom "darlings of the stage" wearing "tights."

Watching over everything was the local government. Even most of the camps organized a loose-knit government and passed a few ordinances. Some of the large towns had fairly well-organized city governments. As in the earlier mining districts, though, most residents did not want the heavy tax burden that often accompanied a mayor, a town council, other officials, and a host of ordinances, so they got by with as few of these structural elements as possible.

Ordinances were necessary, though, and served very practical purposes. For example, Ward, in the mountains in western Boulder County, passed an ordinance regulating dogs, which included a tag requirement and defined "vicious dogs." (Mining communities were often overrun with half-wild "curs and perps [puppies].") Offenses and misdemeanors were carefully spelled out, as were fines. There were curfews for children under the age of sixteen and fines for parents who allowed them to loiter or to be found in "public places" after the appointed hour. To raise revenues, the town passed a schedule of business license fees and used the proceeds of a poll tax to improve the roads. The fire company was volunteer, but was under the "subject and control" of the Board of Trustees. To further address the everpresent concern of fire, an ordinance defined where gunpowder, dynamite, kerosene, and similar materials could be stored or kept within the city limits. (Ordinances did little good, however; almost all camps and towns were ravaged at one time or another by the "fire fiend.")

Unlike Ward, much larger Telluride had a more elaborate city government and set of ordinances. For example, in addition to a mayor, board, and a few officials, it had a street supervisor, health officer, city attorney, cemetery sexton, and a superintendent of water works—all salaried positions. Telluride had much higher fees, a larger budget, and a salaried fire department. The town was also concerned with what today would be called environmental issues, such as a clean river and drinking water and the absence of litter. Its telephone franchise, joined by an electric franchise, indicated a prospering, progressive community. Thriving new mining towns, which did not have to worry about established, older franchises, often quickly accepted modern improvements, especially in communications and transportation.

City governments faced a wide range of issues and problems. Victor, for instance, fought with Colorado Springs over water rights, ordered property owners to number their houses, and faced a financial emergency in June 1913. The treasurer announced that $800 remained in the treasury, and the city needed $2,000 to pay its bills. In its wisdom, the council declared that "a financial emergency exists": to resolve it, they moved to borrow "$2,000 from the Bank of Victor." They supported the Miners' Band with $25 for weekly concerts during the summer, donated money for flowers for Decoration Day and $500 for the July Fourth celebration, and purchased a pest house "to take care of small pox cases." Poorly maintained streets upset the citizens, as did the "disgraceful sights and scenes occurring on Portland Ave. daily."

Georgetown drew upon New England for inspiration for its local government. A Board of Selectmen with no mayor was proposed in the city charter. Limited "executive" power was granted to a justice of the peace, or police judge, who ran the board meetings, dealt with town affairs, and sat as a judge on town matters. This decision combined executive and judicial power in the person of one individual and gave that person an unusual amount of power.

As communities declined, city governments declined with them. The budget for St. Elmo dropped from $2,200 in 1912 to $400 in 1916. License fees were reduced for the few remaining businesses, salaries were reduced, trustees resigned, and trustees' meetings were postponed or regularly skipped. Even when they held meetings, they accomplished little beyond routine matters. Finally, in 1916, a husband and wife became the town's clerk and mayor, respectively. St. Elmo seemed to get along very well with little government during both its brief heyday and its waning days.[5]

City fathers had to come to grips with all aspects of their communities, including the seamy side. Red-light districts, with their saloons, gambling halls, dance halls, parlor houses, cribs, and low-class variety theaters, tended to attract attention, both positive and negative. Reformers railed against the districts and their "hells"; tourists gawked, and maybe even sampled the "sin" offered there, but the districts were a vital part of the male-dominated and -oriented mining communities. Also, there was practicality: If a town or camp did not furnish such entertainment, business would go elsewhere.

Without question, these shady districts both fascinated and repelled people. Mabel Barbee Lee, who grew up in Cripple Creek, retained this memory of her youth about the beautiful Pearl De Vere:

> She called herself Pearl De Vere but even I, an eleven-year-old schoolgirl, knew that wasn't her right name. . . . Once in a breathless instant when I was staring at her from around a telegraph pole, she glanced down and smiled at me. I was spellbound; never had I laid eyes on such an enchanting vision! From that day, Pearl De Vere became my secret sorrow, the heroine of my fondest daydreams, mysterious, fascinating and forbidden. . . . I was more interested in what went on at the Old Homestead [parlor house], and speculated endless[ly] about it. But my rangy musing got me nowhere and I was afraid to ask questions of grownups.[6]

In this respect, too, the towns outdistanced their camp contemporaries—none more openly than Leadville, which flaunted its red-light attractions and seemed to enjoy comparing itself to some of the famous districts back east, such as New York's Tenderloin. The 120 saloons mentioned earlier were joined by 118 gambling houses and private clubrooms, 19 beer halls, and 35 houses of prostitution. That probably stands as a Colorado record! Camps, in contrast, might have a saloon or two and perhaps, in the summer, a "traveling" pair of "erring sisters."

Leadville was an exception to the normal pattern of tolerating but not mentioning this aspect of community life. *Leslie's Illustrated* reported: "Leadville is very wicked; vice in its most civilized and most repugnant form is everywhere rampant." Another report described "Gambling Hells," which flourished because "drinking and gaming are usually the invariable occupations of these men when not engaged in the mines." Saloons "are filled night and day by eager players and drinkers" and "double shifts of dealers and bar keepers are required to meet the ceaseless demand." Said one shocked and scandalized visitor, "In the evening we saw Leadville by gaslight—an awful spectacle of low vice."[7]

If possible, the red-light district was segregated into one part of town so that the problems—drugs, drinking, fighting, crime, prostitution, gambling—would not overflow into the rest of the town. Georgetown's Board of Selectmen, for example, had a long discussion in January 1881 about whether the marshal should enforce the city ordinance and close down the bawdy houses, or just keep such activities from infiltrating into the rest of the town. Ouray's Dave Day commented sarcastically, "The reckless manner in which the soiled doves settle around in various portions of our once moral village should at least suggest a herd law. Let the evil be concentrated or bounced" (February 16, 1883).

Two concerns intersected in this discussion: the obvious morality issue and the less publicly stated economic concern. With ordinances against gambling and the "brides of the multitude" on the books, when they flourished revenue was generated monthly through fines collected. The "sin tax" helped keep other taxes lower.

Although Aspen newspapers seldom mentioned prostitution, the *Aspen Times* of June 18, 1885, observed during a state census that the "sporting women are adverse to telling their real names." The article also noted, "This is not strange as the relations of many of them are under the impression that they are governesses, or milliners, or dressmakers, or music teachers, and that they occupy way-up positions in the society of the Rocky Mountains."

This particular "social evil" generated editorials on various topics, including the influence on children, and "creating a morbid appetite for [a] still higher degree of sensational reading." The Georgetown *Daily Miner* (January 27, 1873) talked about the double standard involved: "The taint of pollution on the character of woman lasts forever, but taint of pollution on the character of man is forgotten or varnished over by an indulgent public—too often by anxious mammas—under the innocent and harmless name of 'sowing wild oats.'"

Whether tarred or tolerated, red-light districts flourished during boom times. When the "drifting crowd" began to move on, the area had a warning that the flush days were waning. As the new century opened, only the towns retained sizable "sin" districts, and even they came under constant fire from churches, women, a few politicians, and even some men.

Gambling was popular in the mining communities, no matter how many might carp about it. Cards, roulette, lotteries, shell games, local baseball games, and almost anything else that could be bet on all flourished. In a sense, miners gambled with their lives every day in the mines, so it was not unusual that they would gamble during their leisure time.

Women were, of course, an integral part of the red-light districts, but few women were included in the initial rushes into Colorado mining districts. To illustrate, in January 1879, Leadville had, among the thousands who rushed there, "about three hundred women in the place, two hundred married and living with their husbands, and one hundred not married, but ought to be." As the years went by, however, more and more women settled in the communities and raised families. More established communities offered more amenities and stability; in turn, the presence of women enhanced the stability and permanence of a town.

Life was not easy. Women found it challenging to maintain a home with an often-transitory miner husband or in a newly birthed, rough mining community. Nor was it simple to keep one's children from seeing or venturing into the seamy side of life. The best way to understand women's experiences while living in a mining community is to let them speak for themselves. In fact, some of the best firsthand accounts of life in the Colorado mining era come from the pens of women, who had more time to write down their daily activities, joys, sorrows, and opinions than did their miner husbands.

Mollie Sanford, who arrived in Gold Hill in July 1861, described her cabin: "There is a rough log cabin, neither chinked nor daubed, as they call it, no floor, and only a hole cut out for a door and window. A 'bunk' is made in one corner. This is covered with pine boughs, and on this are spread our comforters and blankets." When she settled in a cabin with "glass windows and doors," she said "[w]e left our prison in the gulch." Like others in the New West, she became homesick: "I believe one reason for my homesickness is that I do not hear from home. They must have written, but our mails come every way, or *any* way."

Harriet Backus went with her husband to the Tomboy Mine, high above Telluride, and wrote an honest and humorous account of her trials and tribulations trying to cook at 11,500 feet:

> I was one bride who couldn't boil an egg. Only after repeated trials were our frozen eggs boiled long enough to be palatable. It was hard for me to realize that water boiled at only 190 degrees. So for two whole days I boiled beans. They neither swelled nor softened but remained as hard as marbles.
>
> Most of my days were spent perusing the *Rocky Mountain Cook Book,* mixing recipes, washing dishes, in a desperate endeavor to set before my husband appetizing meals from our limited larder. I learned to cook the hard way, but it took time.

Her husband, George, did not think Harriet should have their first child at the Tomboy Mine, so he found her a temporary home near Telluride's hospital.

> I was uncomfortable and found it difficult to walk on our rough paths. My long, brown maternity dress had a short yoke from which the very full gathered skirt hung to the floor. But it would have been highly indelicate not to have such a voluminous disguise to hide my figure. Mother had always said so.
>
> George's birthday was September 2nd and I thought it would be fine if I could give George a living birthday present. So a couple of days before September 2nd I started to walk up the steep paths nearby. It happened that our doctor was in that area and asked what I was climbing like that for. My answer was I wanted the baby to come on the second, but she didn't.

Mabel Barbee Lee recounted her mother's experiences during the start of the Cripple Creek boom when they moved into a tent:

> [Our tent] was patched and dirty, but differed from [others] by having a store door with an adjoining half window. A lumpy mattress sagged on an old wooden bed in a corner and a thin pad covered the cot in the lean-to which I proudly called "my room."
>
> Kitty [her mother] began at once to clean and scour. My father and I . . . made trips to Roberts' Grocery for food supplies. There was little to choose from but Kitty could make the plainest things taste good.

Anne Ellis saw life from the underside, both as a miner's child and later as a miner's wife. In Bonanza, her family consisted of five people crowded together in close quarters:

> Now comes a heavy snow, and we are all penned up in one small room, each time the door is opened the wind driving the snow in and across the floor. Around the tiny stove, German socks and overshoes are drying; across one corner is stretched a rope; this also holds clothes left to dry.

In Cripple Creek, she and her two children heard sad news, the tragedy that had to be in the back of the mind of every miner's wife day in and day out:

> "Well, Mam, you see he drilled into a missed hole, and here we didn't know he had a family. But—well, Mam, you might as well know. He is dead, shot all to pieces."
>
> I just melt on the floor, the quilt covering me. One of the men say, "Look, boys, and see if there is any whiskey or camfire," and tried to raise me, but I, dry-eyed and voiced, ask them to leave me alone for a while.[8]

These women, and thousands like them, followed their husbands along the Colorado mining trails. They turned shacks into livable dwellings, cooked meals under conditions that would appall a modern housewife, and raised their families in the best Victorian manner they could manage under very trying circumstances.

Children grew up fast in mining communities. Young girls, once they became teenagers, were available to be courted in a world very short of eligible women. By that age, boys might be working to help their families as well, although they generally did not do underground mining until they turned sixteen.

Mildred Ekman remembered growing up at the Tomboy Mine. "Kids," she recalled, could find fun such as going "up on the hill and throwing rocks down an old shaft. And listening for them to hit bottom or wherever they hit." Rocks once got Ekman and some friends in trouble. They "liked pretty rocks," picked some up, and piled them behind her house to play with. One day the mine manager appeared and said that "someone [had] reported my dad highgrading behind the house. Mom took him [there]—he laughed and said let them play."

Martha Gibbs recalled that she spent "many happy hours making bouquets of clover blossoms," enjoyed sliding in the winter, liked "walking on top of the board fence around the house," and had fun playing hopscotch at Telluride's school. Ernie Hoffman vividly remembered his Silverton days: "If I got in trouble in school I automatically got in trouble at home, no question about that." Despite that, he "thought a hell of a lot of all my teachers—couldn't say a bad word about any of them."[9]

Most of the teachers were women, with men serving as the principals and teaching in communities that had high schools. A single woman, as Annie Laurie Paddock explained about Creede, "was expected to follow certain standards." Those standards included teaching Sunday School, not being seen alone with a man, wearing "proper clothes," and "never, never" going into a saloon or being seen drinking.

In some senses, women found greater freedom in mining communities than they could have in the more restricted eastern society. For example, more jobs and positions were open to them because so many men were tied up with mining and did not have the time to work in town. They ran stores, taught school (a popular profession for women everywhere), and operated post offices, restaurants, and boardinghouses. A few were even found in traditional male occupations like packing, freighting, and gambling. According to popular ideas of the time, women's voices sounded better over the telephone and their fingers were more fit for typing than men's, so telephone operators and clerks were often women. Divorce, too, did not carry the same social stigma for Colorado women that it did elsewhere in America. In an overwhelmingly male-populated world, men were always eager to have a woman's companionship.

Despite these freedoms, the Victorian ideal of the woman's place being in the home raising the children was an axiom of social organization. Even with upper-class women on the march as the century turned, women in the mining camps tended to be conservative in many ways. They did, however, help orchestrate the 1893 women's suffrage victory, putting Colorado in the forefront of the movement.

They were also in the forefront of efforts to civilize their camps or towns. The presence of a school and a church symbolized urban America. No one supported the establishment of these two institutions more than women. Mining communities were proud of their schools and churches (brick or stone structures were particularly impressive), regardless of whether they financially supported or attended them, because such institutions were signs of permanent settlement and lent at least the appearance of refinement to isolated Colorado mountain outposts. Local newspapers kept an eye on young scholars and their teachers. The editor of the *Rocky Mountain Sun* was shocked during a visit to Aspen's school, reporting that the low number of students would "convince anyone that a number of children were not attending school. The cause was probably lack of interest by their parents, not poverty." He went on to warn, in an April 5, 1890, exposé, "ignorance is a vice which should not be tolerated."

Depending on a variety of circumstances, the school year might last only a few months, or classes might be held in the summer. The equipment available, the subjects taught, and the teachers' qualifications varied considerably. School boards often had a difficult time keeping single female teachers. In this male-dominated, Victorian world, married teachers were generally not approved, and a teacher who was in the "family way" was not at all acceptable.

Meanwhile, students might or might not attend classes, for various reasons. Many families needed more income, and few jobs required much schooling. In some areas, the prevailing feeling was that, having mastered reading and writing and perhaps a smattering of arithmetic, a child was well prepared to venture out in the world. High schools were available in only a few of the towns, and college was well beyond both the means and the aspirations of most members of that generation. Nevertheless, the school was generally the center of community life, particularly in the camps. Social events, graduations, school programs, community Christmas celebrations, plays, and even political rallies were all held at the schools.

In the spring, a young man's fancy might turn to that cute young lady in his class—but more likely, it turned to baseball. Baseball was wildly popular in mining communities large and small. Every town and a number of camps fielded a "nine" that played to uphold the community's reputation and garner the money locals bet on their favorite team.

The *Rocky Mountain Sun* (September 12, 1885) reported that Crested Butte "is itching for a game with the Aspen nine. Why don't they come over here, it is their turn to cross the range anyway." In 1871, Central City accused Denver of gathering players "from the four-quarters of the globe, veterans of years standing." Rivalries between towns often grew, particularly if a team sported some "ringers." The local paper might complain, as did Silver Plume's *Silver Standard* (May 29, 1886), if the home nine had a victory "stolen" from them.

Familiar cries that "the umpiring was rotten" and "our boys batted out of luck" graced many reports of games. The *Creede Candle* (July 27, 1907) became particularly aggrieved about the umpire at Monte Vista, "who rankly favored his home team." It concluded: "Umpires who allow prejudice to supplant honor and honesty will eventually kill baseball if tolerated." Sadly, visiting teams often faced this problem.

The greatest Colorado team of the nineteenth century, the Leadville Blues, took to the diamond in 1882. Determined to overcome its 1880 image of mining failures, the Cloud City fielded a team made up of five future or former Major Leaguers. They dominated the Colorado scene, defeating teams everywhere from Denver to small mining camps, often by large scores. Then they took to the road, but got their comeuppance when they ran into a team in Council Bluffs made up of Chicago players from the recently completed National League season. The Blues took the field in other seasons, but never again with such success.

Scores were high in those early decades, with poor playing fields and no gloves—gentlemen would not wear them. Though the Central City Stars were "woefully defeated," 55–30, the local paper still had hope: "We are not disposed to give up our faith in them. We firmly believe they will yet be so developed as to then enable them to meet and overcome the best club in the Territory." Locals followed their teams with pride and occasional dismay at an "overwhelming defeat."

These were the decades, as the *Rocky Mountain News* proclaimed in 1867, when baseball was "the American national sport, and [no other] outdoor amusement or exercise equals it in genuine healthful recreation." The reporter continued, "As a means of invigorating manhood and developing the physical powers of youth, it stands without a rival." Long before Casey strode to the plate, mining-camp newspapers published baseball poems. Even after the turn of the century, springtime brought forth a poem or two:

> In the spring the young girl's fancy
> lightly turns to thoughts of hat, . . .
> In the spring the young man's fancy
> lightly turns to thoughts of ball.[10]

For all its popularity, baseball did not dominate the sporting scene entirely. Boxing, wrestling, bicycling, running, shooting matches, fire company races, and eventually basketball and football all attracted fans. Even cricket was played a few times. These events attracted men more than women, and betting was a constant and major adjunct to all events.

Women were more likely to be involved in the local church. In fact, they were the backbone and mainstays of mining-community churches. Church was an acceptable social outlet for both women and their children in these predominantly male enclaves. Further, women could play a leadership role in churches that might be denied to them elsewhere. The Reverend James Gibbons, who served Catholic churches throughout Colorado mining towns, said this about women:

> As workers in the church, they were second to none. The women attended not only to the proper duties of the altar society, but in no

> [small] measure to the financial affairs of the church. Fairs and balls were organized and managed by them, the tickets sold, the collections made and the money put in the bank to the credit of the church.

Churches needed that support, because Sunday was a wide-open business and social day on which various entertainments and activities (some highly secular) competed with worship.

It took a special type of minister or priest to match his congregation's needs. Presbyterian George Darley, who served throughout the San Juans and elsewhere in Colorado, felt that in "no region" can men be found who "are so indifferent to religious influence as in a new mining camp." Yet, he opined that there "are more faithful followers of Christ than some men like to have us believe."

The materialism of the mining communities challenged religion at every turn. Darley felt you "had to suffer and suffer again" to minister to these people. Physically, it was demanding. Darley walked 125 miles, "more than half the distance through deep snow," in five days and four nights to reach Ouray. Years later, in writing of the difficulties he and other pioneer ministers faced, he concluded that the obstacles "were more than the average minister cared to face then or would care to face now."[11]

Darley met people where they were, not needing a church and a pulpit. He preached in saloons, homes, and other places that might have shocked the faithful back East. A "poor" minister did not last long, nor did one steeped in unbending conservatism or one whose ultimate goal was simply to build a "house of God." Few mining camps had enough Methodists, Baptists, or any other denomination to corral the members needed to maintain a church. Ecumenism was the answer, at least among the Protestants.

The church and school were the centers of respectable entertainment. Women of the former might sponsor dinners to help raise money for building or some other church need. These dinners were popular with single males, of which every community had an abundance. School functions allowed proud parents to watch their sons and daughters show off their talents at music or oratory.

One way to overcome the transitory and isolating nature of mining work was to join a fraternal lodge, some of which also had women's auxiliaries and children's groups. The Masons, Woodmen of the World, Elks, and other lodges provided entry into a new community for members; they also sponsored dances and other social events along with regular meetings. For Union veterans, the Grand Army of the Republic held meetings, sponsored social events, and marched in parades; it also worked hard to get pensions for its steadily aging members. Every two or four years, the two major political parties, and some of the minor ones, organized "clubs" to advance their candidates.

Americans and Coloradans in those years were joiners. The era represented the high tide of fraternal orders and clubs such as literary groups and Chautauqua meetings with its social, educational, and cultural events. Mining communities did not lag behind in organizing, participating in, and enjoying clubs. As the largest of Colorado's mining communities, Leadville offered the greatest variety of clubs and societies, including (among others) the Bible Society, Band, Athletic Club, Glee Club, Women's Club, Temperance Club, Women's Christian Temperance Union, Y.M.C.A., Pioneers Society, Jockey Club, and Library Association. If one cared to belong, there was an organization available; no one need be left out. There were even enough blacks in the community to allow them to organize both social and political clubs.

In the 1890s and the first decade of the twentieth century, the Western Federation of Miners became very active in working for better pay, safer working conditions, and members' rights. The union also helped build hospitals in several San Juan communities. In the antagonistic labor world of turn-of-the-century America, however, unions fought a losing battle, as will be discussed in the next two chapters.

Colorado mining communities reflected the trends of the United States overall, as did the neighboring farm and ranching towns. The bicycle craze hit in the 1890s, just as in the Midwest and parts East. The safety bicycle, which replaced the large front-wheeled bike (the "ordinary") of previous years, found ready acceptance. People rode bikes all over the mountains on roads that were hardly more than trails. Local newspapers reported their adventures and hailed the arrival of bicycle salesmen as a harbinger of spring. Bicycle clubs were formed, and they, too, sponsored social occasions. The ladies rode, too, although they were careful to wear long dresses, or maybe "bloomers" that would cover their sexy ankles.

Interestingly, as single- and double-jack equipment was replaced by machine drills in the mines, hand-drilling contests gained popularity. On holidays, such as the Fourth of July or Labor Day, a baseball game and drilling contest often highlighted the occasion. An individual or a team competed in a timed contest to see who could drill the deepest into a granite boulder. Prizes of a hundred or more dollars were not uncommon—compared to miners' wages, a profitable few minutes' work—and "champions" were hailed in the press and often recognized regionally.

The 1893 Fourth of July celebration in Creede was typical. The double-jack contest offered $250 for first place and the single-jack prize was $50; the winning baseball team received $300. There were "fast and slow" bicycle and burro races, plus contests purely for entertainment, such as the blindfolded wheelbarrow race and the sack race. A bowling contest was also included.

The other major holiday season stretched from Christmas through New Year's, when the mines closed and the miners came to town. Fraternal organizations and clubs sponsored dances, the churches held various celebrations, and the schools often displayed their young scholars singing and reciting holiday carols and verses. Santa Claus might also make an appearance with candy for the "good boys and girls." The *Creede Candle*, December 23, 1892, wished: "May Santa Claus, he of the white beard, jolly face, rotund figure and the reindeers, fill each and every one of your socks clear to the brim with good things."

New Year's Eve might have a few "watch night" services, but was generally given over to fun and frolic. Cripple Creek, at the height of its boom, welcomed the new year and the new century in this fashion, according to the *Star*. "At the stroke of 12 not only did pandemonium break loose in the way of shrieking whistles and booming dynamite and yelling people but [also in] a flow of free ardent and malt beverages at the different saloons." The reporter added that in spite of this commotion, there was "very little drunkenness observed" and "no disturbances."

The modern world entered mining communities in many ways, both in private homes and on Main Street. Electricity promised numerous benefits for Colorado mining and mining towns, and in the 1890s it spread throughout the mountains. Two types of electrical power were being experimented with, direct current and alternating current. In theory, alternating current was better, because higher voltages could be transported more cheaply over long distances; however, the direct-current process championed by Edison on the eastern seaboard was adopted by many communities and industries, including Aspen. Its main problem was loss of voltage over distance; power stations had to be built about every two miles.

In the mountains, alternating current seemed more advantageous, once the theory evolved into practical reality. It did so thanks to Lucien L. Nunn and his Gold King Mine high in the San Juans, which needed an economical power source. As a result, the nation's first alternating-current power plant opened in 1891 at Ames near Telluride. Running six days a week, the plant ushered in a new era, as well as providing quite a show for intrigued onlookers when the sparking, hissing generators were started. In the years that followed, electricity replaced older power and illumination sources, such as steam and gas, wherever possible. It forever changed work in the mines and life in the mining communities.

The automobile, that chugging monster that scared horses, made its appearance after the turn of the century. The "horseless carriage" reached Leadville via Colorado Springs in 1902, a fairly easy, day-long drive. By comparison, the first car to travel over the well-named Stoney Pass into Silverton took five days in 1910 and even then made it only with the help of horses.

Speeding autos soon became a problem, and city councils had to devise acceptable speed limits and prohibit such activities as driving on "sidewalks." Because the automobile was new, it was a learning experience for all. Telluride handled the problem by prohibiting anyone from going faster than eight miles per hour on the main streets or from going through an intersection or turning a corner faster than four mph. In the rest of the town, drivers could go ten mph. If they struck "any pedestrian or another vehicle," however, the fine ranged from $10 to $100.

The "auto craze" led an increased demand for better roads, although the clamor actually started with the bicyclists, who soon tired of bumping over rocks and bogging down in seemingly bottomless mud holes. It was no better for the early auto drivers. For example, Creede cheered the arrival of five "automobile loads" of tourists in August 1914, all the way from Kansas City. The only problem they noticed, reported the *Creede Candle* (August 8, 1914) was "muddy roads," although not in Mineral County, which they praised "liberally."

Part of the problem was the lack of state money for roads. The funds that were available tended to go to the more populated regions along the foothills. Cash-strapped mining counties could not offer a great deal of assistance, as much as they might have wanted to. Mountain roads also posed unusual engineering and maintenance issues, which only became worse in the high mountain mining districts. One result of these problems was the Good Roads Association, which promoted both better roads and tourism. It sponsored the Good Roads Day, when locals worked on various sections of nearby roads filling potholes, putting up road signs, grading surfaces, and doing anything else that might help.

The new transportation wonder benefited mining in more ways than the car. Trucks offered almost unlimited potential for hauling ore, supplies, and other goods more cheaply, conveniently, and quickly than mule and burro trains—or even the railroad, for that matter. The railroads that had long been essential to mining began to decline in importance.

Leadville, as well as other mining towns, saw the possibilities the auto offered. Georgetown, Aspen, Gothic, Ouray—towns large and small—hoped that tourists would come for a "delightful summer" stay in the "cool, beautiful" mountains that surrounded them. In May 1913, Leadville formed a Commercial Club to publicize Lake County and attract "motor tourists." The *Herald Democrat* proclaimed: "The automobile is doing wonders in linking the whole country together. It is doing this in a more intimate sense, even, than the railways." That attitude soon put an end to railroads' domination of tourism, because "the leisurely auto tourist makes a point to stop a while at various points of interest."

Promote or die! Mining communities promoted tourism to the best of their limited financial abilities. They joined associations, published articles in magazines, wrote feature stories for local newspapers, and encouraged locals to "be nice to visitors." This included, for Georgetown, having comfortable rooms available for tourists and sprucing up the town to make a favorable impression on first-time visitors. They also encouraged railroad travel bureaus to promote stopping for a visit, whether for sightseeing, fishing, or simply relaxation. A few communities that had or were near hot springs, such as Ouray and Idaho Springs, promoted the health-giving qualities of their mineral waters, whether by drinking or enjoying a "life-restoring" bath.

More than scenery drew tourists, however. Some came to see history before it slipped away. In a few still-prosperous towns, like Creede in the 1890s or Telluride and Cripple Creek just past the turn of the century, they could see the "real McCoy," with all its attractions and warts—the "vanished" West was right there before their eyes. More likely, they might wander around ghost towns, trying to imagine what had transpired there and talking to old-timers who had been there when times were flush. These old-timers actually became a major Colorado tourist lure: They would sit in the warm sun on summer days and reminisce (perhaps remembering more than actually happened) with visitors about times and people long gone.

Even as they spun their yarns, a new invention was changing the image of the West forever: the motion picture. The image, though, was not of mining but of the legendary "Gunsmoke and Gallop West." As the twentieth century dawned and matured, opera houses like the Tabor in Leadville became movie theaters, as did old stores or any vacant buildings that could be modified to show a movie. Initially, one-reel films, then two- and three-reelers soon appeared in "theaters" in the towns and even some of the camps. The *Silverton Weekly Miner* (August 7, 1914), promoted *When the Earth Trembled,* called "pathetic and thrilling" and "one of the best moving picture shows."

But it was the Western that gained the most popularity, starting with the very first one, *The Great Train Robbery*. Mining never became a staple subject for Hollywood.

As the United States entered World War I and the Colorado mining camps and towns slipped into history, legend, and folklore, few stopped to look back. A new era had dawned, a new twentieth-century Colorado that looked ahead rather than to the past. That teenager of the Pike's Peak rush, now well into his or her seventies, would have witnessed an amazing transformation. At the same time, he or she might look back longingly at the days of youth and mining's high time. A song from that Victorian era, "After the Ball," captured melancholy remembrances of a time that now seemed quaint:

> After the ball is over,
> After the break of morn,
> After the dancers' leaving,
> After the stars are gone;
> Many a heart is aching,
> If you could read them;
> Many the hopes that have vanish'd
> After the ball.

9

"The Everlasting Love of the Game"

Mabel Barbee Lee, in trying to portray the allure of mining for her readers, recalled a meeting she had had with an old-time Cripple Creek prospector during a 1951 visit to the then ghost-like town. No longer the exciting, booming "metropolis" of her youth, Cripple Creek languished in yesterday:

> The shine of hope and faith in the old fellow's eyes followed me long after he had disappeared from sight, and it came to me, as it had once long ago, that it wasn't the gold he wanted. It would likely slip through his fingers in no time, or be given away for the asking. It was the enticing hunt that led him on, the elusive chase, the everlasting love of the game.

The "enticing hunt," the "elusive chase," the "everlasting love," or, as David Lavender described it in his novel, *Red Mountain*, "the frame of mind of the people"[1]—all these things spurred the miners on as they rushed, claimed, developed, and then moved again to chase a new dream.

It had started with the Pike's Peak rush of 1859, which even in the 1890s seemed a long time ago. By 1890, a generation had passed and the early days had become the stuff of history and romanticized stories. Industrial mining had replaced the legendary prospector and his burro, and workaday reality had supplanted the strike-it-rich dreams of yesteryear. Except for Aspen, there had been little to cheer about recently. Colorado mining seemed to have reached old age as the century neared its end. Mining would continue, but it lacked the glamor, the excitement, and the individual rags-to-riches stories of wealth that spawned early Colorado mining history legends And yet, in that last decade of the nineteenth century, a faith—a fond hope—still tantalizingly beckoned: that maybe, somewhere over the next mountain, the "mother lode" still lay hidden waiting for some lucky prospector.

The 1890s, the "Gay Nineties" of folklore and legend, were anything but gay in Colorado mining. There were a few exciting days, but there were far more depressing ones. Before the decade ended, the "jack ass" prospector's dream of stumbling into his own private bonanza would be gone, and the day miner's hope of owning his own mine would fade away. In their place would be the "modern" industrial world of absentee owners, professional management, stockholder demands, and labor/management confrontations.

Silver did produce one more excitement early in 1891, when Creede roared into the spotlight and reached the $1 million production level within a year. "It was day all day in the daytime, and there is no night in Creede," sang local newspaperman-poet Cy Warman. Like its fellow strike areas, Creede, in its own mind, was the best-ever mining district. Look out, world, shouted the *Creede Candle* (January 14, 1892): "Certain it is, no other camp yet opened in Colorado could show six as rich producing mines as those opened within eight months from the time the first real intelligent prospecting was done." Overpromoted as a second Leadville and overly exploited, Creede shone briefly, but fate dealt it a bad hand. It was silver's last hurrah.

Except for the continuing decline in the price of silver, the early years of the decade were good to the state's mining industry. With silver production at about the $20 million level and gold moving up from $4 million to $5 million annually, Colorado mining had never produced so much. To maintain that level of silver production, however, more ore had to be mined every year, taking from the future to pay for the present.

Part of the euphoria resulted from the passage of the Sherman Silver Purchase Act in 1890. After more than a decade of wrangling, the silverites finally caught a break. Eastern Republicans were terrified by the possibility of inflation, which they were convinced "free silver" would bring about, but they needed western votes to pass a high protectionist tariff. Westerners were not strongly in favor of the idea, but were willing to go along if a bargain could be struck. The resulting vote-trading led to passage of the Sherman Silver Purchase Act, balanced by the higher McKinley tariff. After enactment of the former, the Treasury was required to purchase 4.5 million ounces of silver per month, or the assumed total U.S. production of 54 million ounces per year. One of the main differences between this act and the earlier Bland-Allison Act was the substitution of ounces for dollars'-worth purchased.

To placate unhappy "gold bugs," the silver certificates issued could be redeemed in either gold or silver at the discretion of the Secretary of the Treasury. Western silver interests had again gained only half a loaf. More silver would be purchased, but there was no guaranteed price or price ratio, such as the long-sought 16-to-1 silver-to-gold ratio that Easterners stubbornly refused to grant. Such horse-trading satisfied no one completely. Conservative Easterners still wanted what to them was "sacred": the gold standard and "sound money." Eastern creditors also feared that the plan was a concession to "cheap money," foreseeing that debts would be repaid with money worth less than the amounts they had loaned to debtor Westerners.

For a while, though, the plan seemed to work. The price of silver jumped to $1.07 an ounce in 1890, but then fell steadily until it reached a dismal 78 cents an ounce in 1893 (roughly a 36 percent drop). Again the silverites rose in angry protest, but now they had a much broader support base.

In the simplest terms, farmers were having nearly the same troubles as the miners, but for different reasons. Postwar expansion onto the Great Plains and beyond had greatly increased the number of farmers, who, thanks to virgin land and new seed varieties, improved farming methods, and better equipment, were able to produce more crops—but at a cost. To finance land and everything else they needed, they had to borrow money, mostly from eastern bankers and other investors. The market did not expand as rapidly as production, and the resulting surplus lowered prices. However, the prices the farmers paid for equipment and the high interest on the funds they had borrowed did not decline.

The farmer—once the "backbone of America"—found himself marginalized, in debt, passed over by prosperity, and derided as a "hick" and a "hayseed" by his more "up-to-date" urban cousins. Agrarian spokespersons thought there was "a screw loose" in the American economic system, and from their perspective indeed there was. Farmer protest groups that appeared in the 1870s and 1880s finally coalesced into the People's Party, better known as the Populist Party. Farmers and miners had similar grievances—eastern creditors, eastern bankers, unsympathetic Washington, big business, unresponsive Republican and Democratic Parties, and hard times spreading—and thus they decided to run their own candidates on their own issues.

"All power to the people," not to entrenched interests, became an appealing rallying cry. Both groups wanted "cheap money," which they believed would produce inflation and raise the prices of silver and crops. Shocked eastern conservatives recoiled, horrified at such heresy, such radicalism.

Silverites in Colorado enthusiastically cheered when the Populists announced, among their other 1892 platform positions, "the free and unlimited coinage of silver at the ratio of 16 ounces of silver to 1 of gold." The "silver issue" took on intensified meaning. It was free silver, and free silver alone, that Coloradans wanted. The Populists' and other parties' ideas might be fine, and some might actually benefit Colorado, but free silver directly championed their interests and their future.

Silver was going to help the miners, farming folk, city dwellers, debtors, and anyone else who believed. Belief was the key. This particular political/economic stance incorporated a strange concoction of emotion and desperation, logic and illogic, hope and fear, confidence in the cause and dismay that others could not see the justice of their stand, all based on the fact that the American dream had become a nightmare. It was rural America vs. urban America, old America vs. new America, the common folk vs. the rich, us vs. them—and it was cast in terms that were part carney pitch and part hardheaded reality.

The Populists, less a party than a cause, were "marching to Zion" with a near-religious fervor. Coloradans certainly believed in Populism in 1892. They gave their presidential electoral votes to Populist James Weaver, elected the outspoken Populist Davis Waite governor, and sent Lafe Pence to the United States House of Representatives. Populists also gained control of Colorado's new House seat with the election of John Bell. With the aid of the Democrats, they also controlled the state senate and came within one vote of a house majority.

With revolution brewing at home, Colorado's two Republican senators, Henry Teller and Edward Wolcott, worked hard to convince their congressional colleagues of the correctness of the silverites' position. Joined by "silver" congressmen from Montana, Nevada, and Idaho, they spoke, lobbied, advocated, wrote, and tried in every manner imaginable to advance the silver cause.

Teller worked endlessly at promoting the idea of making silver and gold equal in the United States monetary world—a ploy that, it was hoped, would make silver worth $1.25 an ounce. On January 6, 1892, he rose to give a long, detailed speech in the Senate. Though not a spell-binding orator, he presented the silver question well, both logically and emotionally. Salient points included:

- Both metals were "indispensable to the prosperity of any nation."
- Silver must be granted "full recognition as a money metal."
- Silver and gold had coexisted throughout history.
- The country needed more money in circulation. "Without a sufficient amount of money with which to do business, the energies of the people are depressed and in time destroyed."
- Every nation "except this" one has hailed its precious-metals production "as a boon from the Almighty."
- If the people were allowed to vote, a "very very decided majority" would be in favor of bimetallism.
- Silver "better meets the wants of the people of this country [than gold], as it does the wants of the human race."
- It is not a question of "cheap money" versus "honest money."

On April 20, he again defended silver to his colleagues, pointing out crucial issues of the times. "We have fallen upon evil times. We have felt the great power, the tremendous influence of political and partisan attachments and political and party relations." Teller concluded: "I may be a fanatic, I may be an enthusiast. Every word I have uttered upon this subject lies close to my heart. I warn my party . . . it can not afford to put itself on the side of a contraction to the extent of one-half of the volume of the money of the world."[2] Teller and his silver colleagues probably convinced few listeners with such speeches, but the people in the silver states applauded their efforts and felt well represented.

That, then, was where the matter stood, when the fateful year 1893 dawned. President Grover Cleveland, who had just begun his second term, soon faced a major crisis. An economic panic and crash quickly became the worst depression the country had ever faced.

The complex causes had both domestic and international origins, but the event that triggered the collapse was the gold reserve falling below the $100 million mark for the first time in the country's history. This created emotional fear that the government might be nearing bankruptcy and might have to suspend gold payments, which created a cascading clamor among eastern businessmen and gold advocates. As the gold reserve continued to dip, Cleveland and his supporters took the position that the Sherman Silver Purchase Act was responsible for the nation's woes, and he called for a special session of Congress to repeal it.

This set the stage for the final struggle between the gold bugs and the silverites, sound moneyists and inflationists, debtors and creditors, East and West—all convinced they had to save the country. Meanwhile, Coloradans were trapped in what might have been the worst depression in the state's history. It spread everywhere, from farm to town to mine; no one seemed immune. The full force hit in July 1893. Within a few days, twelve Denver banks failed, smelters stopped operating, real estate values tumbled, businesses failed, railroads edged toward bankruptcy, and men were thrown out of work. Desperate people traveled around aimlessly, seeking any kind of work.

Press reports from the eastern Plains to the Western Slope reveal the frantic, distressed times. "Shoulder to shoulder, men, while the war upon Colorado continues." "The once flourishing mining camp of Aspen is now badly prostrated by a crushing blow to its great industry. It has been a complete knock-out." "Those who remain in town must take courage, there is a better time coming." "There are many desperate people in all parts of the state at this time."[3] Meanwhile, poor Creede's youth and boom days vanished in a twinkling.

In this era, which predated federal government assistance by several decades, the business community was expected to pull the country out of the depression. It could not do so. Relief groups and churches soon ran out of money, and men and families moved out of the mountains to Denver, frantically searching for work or help of any kind. None was available.

Though it is hard to express in written words the shock, heartache, and bitterness of the disheartening, miserable summer months of 1893, a report from the Colorado Bureau of Labor Statistics came close. The report opened with a typically verbose Victorian sentence:

> From the beginning of the existing financial depression, we have all been more or less impressed with a sense of the wide-spread devastation being wrought in our state through the prostration of an industry which has filled our otherwise solitary mountains with thousands of our bravest and most stalwart citizens, who have built road-ways along the dizzy heights and beetling crags, where the eagle once circled in unbroken silence, amidst the awful grandeur of nature's work in her sternest mood.

Reports from the various mining districts presented the grim picture of the collapse:

Custer County [Silver Cliff, Rosita]: "The people are almost destitute, owing to failure of crops and decreasing price of silver. Money cannot be got on any security."

San Juan County [Silverton]: "Our only hope is a favorable silver legislation, if that is knocked out, we go out with it."

Lake County [Leadville]: "Business is completely stagnated. The people are leaving as rapidly as their friends in other places can send them money to get away with. The approaching winter finds business men and laborers alike standing on the verge of bankruptcy, hunger and destitution."

San Miguel County [Telluride]: "With the Sherman law repealed and no substitute given us favorable to silver[,] 'our name is mud.'"

Summit County [Breckenridge]: "The feeling among the miners interested in silver-lead mines is rather blue."

Rio Grande County [Del Norte]: "This valley feels the prevailing depression heavily because of the falling off in work at mines ordinarily using valley products."

Pitkin County [Aspen]: "The situation is bad and couldn't be very much worse. If we get no favorable legislation, Aspen and vicinity is a goner."

The BLS report also carried dismal news of unemployment and mine closings in mining communities throughout Colorado. A few samples show the drastic impact:

	Unemployed	*Mines Closed*
Aspen	2,000	50
Fairplay	1,000	40
Georgetown	600	45
Leadville	2,500	90
Ouray	1,800	20
Silver Cliff	500	10
Silverton	1,000	10

Even if some of these were merely emotion-based guesses, they told the tale of woe in the bleak silver districts.

Such terms as "gloomy, very bad, depressing, hopeless, serious, blue, dark" summarized the "general feelings" and signaled the collapse of an era and its dreams.[4] Even the gold districts, such as Central City and Cripple Creek, garnered only fair reports at best. Coloradans looked to Washington, where their silver champions—Henry Teller, his junior colleague Edward Wolcott, and their allies—would need to fight the good fight on the Senate floor.

Congress met in August, and the fight over repeal dragged on for three months. The silverites gave impassioned speeches; the sound-money advocates and creditor interests replied in kind. The outnumbered farmer and silver senators filibustered, but it did no good. In the end, the Sherman Silver Purchase Act was repealed, even though doing so solved nothing. The country and Colorado slid deeper into recession, which in Colorado lasted almost the entire decade.

The time of silver in Colorado was over. Many of the silver camps would never recover. In towns such as Georgetown, Aspen, and Leadville, waning prosperity marked the end of an era, as silver production began its long decline. Coloradans might chant—and even believe—that "Silver Is Still King," but its reign had ended when the price of silver plummeted to the fifty-cents-an-ounce range.

The situation in the state would have been even worse if not for the discovery of the Cripple Creek gold bonanza, the greatest in Colorado history. It was the one final, electrifying moment before the era disappeared forever into memory and myth and left it to history to reconstruct what had happened, where, and why. Situated in a volcanic bowl west of that beacon of the 1859 rush, Pike's Peak, the Cripple Creek area had been generally ignored. A brief flurry of excitement over purported gold strikes in 1884, which became known as the Mount Pisgah hoax, ended with the discovery that the claims had been salted; the deception gave the region a long-lasting bad name. What transpired in 1890, however, became a legendary story of a western gold discovery.

Most people assumed that this was cattle country, not mining country. A few prospectors had drifted in and out, but one man, Bob Womack—part-time cowboy, part-time prospector—continued searching for the gold he ardently believed was there. He found some promising gold float and tried to locate its origin higher in the hills, without a great deal of success. Unfortunately, he was an undistinguished fellow, given to carousing when he ventured down to Colorado Springs, and appeared to be only a teller of "tall tales" about his gold. Hence, little credence was attached to his statements.

Finally, in October 1890, after a decade or so of prospecting, Womack found what he had been looking for and staked the El Paso claim. His ore assayed at $250 to the ton. Finally, there was something substantial on the other side of the mountain from Colorado Springs that aroused interest.

Womack's moment of fame was already over, however. He, like so many of his kind, had discovered the wealth for others to enjoy. Bob Womack sold his claim, drank a lot, and in the end benefited little from his discovery except for the celebrity status of having opened Cripple Creek.

In the spring and summer of 1891, the rush was on, with claims staked on nearly every hill in the immediate vicinity of Womack's El Paso discovery. A mining district, organized in April, was named after nearby Cripple Creek, and little camps soon dotted the landscape. Both the tenderfoots and the experienced prospectors/miners who dashed into the district had nearly the same guarantee of success. As the summer wore on, the hills literally swarmed with prospectors, quickly followed by the rest of the denizens who populated a new mining area. Coloradans had seen it all before . . . but something seemed different about this rush.

There had been no other gold district like this in Colorado. Located in a 10,000-acre bowl of volcanic rock that had been fragmented by eruptive explosions, Cripple Creek showed little surface gold. The ore veins, relatively narrow and not marked by quartz outcroppings, were not sufficiently different from surrounding rock to draw attention. Veins could be discovered only by blasting into the ground or sinking a test shaft.

Valuable ore was sometimes thrown out on the dump; conversely, worthless gangue was sometimes shipped in error. Continual assaying, a potentially costly procedure, had to be done to determine the worth of one's claim—then it took capital to open and develop the property. Despite its humble discoverer, Cripple Creek was not the storied "poor man's diggings." The *Colorado Springs Gazette* (June 24, 1891) was correct when it warned readers: "It will take a long time for the district to develop and a good deal of money to find out whether there is gold at Cripple Creek. Cripple Creek is not a poor man's camp."

Nevertheless, one man bucked the trend: Winfield Scott Stratton. Stratton's story was even more classic than Womack's. Stratton had come to Colorado back in the 1870s and caught a bad case of gold fever. A skilled carpenter, he would ply his trade in Colorado Springs in the winter and prospect in the summer. He read books and even enrolled in a mineralogy course at the new Colorado College. He prospected everywhere—in the San Juans and at Leadville, Silver Cliff, Aspen, Red Cliff, and Tin Cup—without noticeable success.

By the spring of 1891, the now-experienced Stratton had prospected and mined in Colorado for seventeen years. Once more he ventured out, this time nearly in his backyard, to Cripple Creek. On July 4, he finally made the discovery he had sought for so long, staking two claims on Battle Mountain, the Independence and the Washington. He was about to become a legend.

Not quite overnight, though. Initially, he thought the Independence was not valuable and gave a thirty-day option on the property to a San Francisco mining group. The night before they took over, however, Stratton uncovered a high-grade gold vein in a crosscut while removing his equipment. He quickly covered it and spent a nerve-wracking thirty days waiting. The mining group found no valuable ore and gave up the option, to Stratton's immense relief. He went back to mining and became a multimillionaire. Conservatively, he held down production; according to legend, limiting himself to $2,000 net per day.

His good fortune overflowed. In January 1892, two Irishmen, Jimmie Burns and Jimmie Doyle, staked a mini-sized claim on Battle Mountain they called the Portland. Located on just 69/1000 of an acre, it was surrounded by other claims. As luck would have it, they hit a rich gold vein, but immediately realized that the adjoining claimants would dispute ownership, based on the contentious apex law. Burns and Doyle approached Stratton, who became their partner, and the three braced for a fight. Cannily, they managed to purchase enough claims to head off potential lawsuits, and eventually the three men ended up owning 183 acres of Battle Mountain. It cost more than a million dollars to buy the land, but Stratton and his partners ended up controlling the richest mines on the richest mountain in the district.

By October 1891, with the first significant ore deliveries to the Denver smelters, Cripple Creek's fame was assured. Even the skeptical *Gazette* seemed convinced. The October 24 issue proclaimed it a "very lively" district, "with buildings going up in every direction." Gold production from the district almost met even the wildest predictions, jumping from $2 million in 1893 to $18 million in the peak year 1900. Once again Colorado became the darling of the mining world. Investors and the rest of the typical mining-rush crowd arrived, along with hordes of unemployed miners from the silver districts throughout the state.

The result was predictable: Colorado's first major labor dispute and strike. There had been one large strike before, at the Chrysolite Mine (see chapter 6), but there had been no strong miners' union. Meanwhile, the hard times had caused owners and companies to cut back wages, release miners, and generally retrench—which had led to small strikes at Aspen, Rico, and Creede over these issues, plus the thorny, emotionally fraught one of unionization itself. The Western Federation of Miners (WFM), which had originated in Butte, Montana, now aimed to organize miners at Cripple Creek.

If it could win in Colorado's most prosperous district, it would gain great momentum to organize miners throughout the state.

Initially, though, the owners held the advantage. Miners flocked to Cripple Creek looking for work. A group of owners moved to reduce wages from the standard $3 to $2.50 per day, willing to risk a strike with so many out-of-work miners clamoring for jobs. The WFM reacted by organizing a local union, and positions on both sides hardened. Unable to agree on a district-wide wage scale and hours, some of the owners took the offensive in January 1894 and unilaterally instituted a nine-hour day at $3, while refusing to deal with the union.

The owners made pious statements about good working conditions and other districts' hours, but the miners went out on strike anyway. They had their own problems, including the fact that a miner could barely maintain a family on $3 a day, even if he worked all the time, which many did not. Even in a depression, Cripple Creek's boom days kept prices higher than elsewhere.

To the amazement of the belligerent owners, the tightly organized miners stayed out into March. The owners' position was also undercut by the fact that neither the Portland nor the Independence mines followed their lead; both continued working. To the chagrin of the other owners, Stratton, the old prospector-miner, refused to cut wages, and continued making money while his counterparts' properties sat idle. He treated his men with more consideration than did many of his contemporaries. Further antagonizing the owners, strikers from their headquarters at Altman, the highest and most unionized camp in the district, defied them too.

As the situation deteriorated, Governor Davis Waite sent in the state militia, which calmed things down for a while, but when the troops departed peace did not ensue. The owners, who controlled the sheriff's office, sent a trainload of recently deputized "bully boys," many from Denver, to teach the miners a lesson. As they passed the Strong mine, on their way to Altman, the union stronghold, the mine buildings blew up, showering the train with flying metal and wood debris; the panicked deputies quickly fled. With the recurrence of violence, Waite sent the guard back in. To the surprise and chagrin of the owners, the guard proved neutral, siding with no one. Finally, after four months, the owners were dragged reluctantly to the bargaining table. Under pressure from Waite, invited as a union representative, they conceded to an eight-hour day at the standard $3 wage.

The WFM had won a strike, one of the longest and fiercest to date. With confidence thus bolstered, it set about to organize all of Colorado. As both sides realized, this was not the end but only the beginning. The fight between management and labor would last two decades, spreading from hard rock to coalfields, before finally ending in tragedy on a spring day in April 1914, at Ludlow, in Colorado's southern coalfield.

Management learned a lesson. It needed to organize into a cohesive unit and elect a supportive governor and, if possible, legislature. With the election of Republican governor Albert McIntire in 1894, the owners gained a valuable ally.

When the WFM organized the Leadville miners and demanded a return to the $3 daily wage, which had been lowered to $2.50 by joint agreement in 1893 to help keep the district alive, the owners were ready. When that demand was refused, the WFM, riding a wave of popularity and confidence after Cripple Creek, called a strike. Both sides dug in. The owners quickly fortified their properties and brought in guards and strikebreakers, or scabs, while refusing to deal with the union representatives.

Meanwhile, with the mines closed, some flooded, an unfortunate occurrence that slowed Leadville mining for the rest of the decade. Violence erupted at mines operated by nonunion miners. In September, a mob attacked the scabs working at Leadville's Coronado and Emmet mines, killing several. The call went out for help, and McIntire sent in the guard—which, this time, sided with management and set out to break the union once and for all. With the guard protecting the nonunion workers, the mines gradually returned to operation, and the strikers could only watch bitterly.

The strike dragged on to a tortuous end, with the miners unable to match the staying power of the owners who were backed by the state. Many miners left, and the rest conceded dejectedly and went back to work on the employers' terms as the strike collapsed. The Leadville union was broken, the WFM suffered a major setback statewide, and the owners learned further lessons they would not soon forget. Neither side was happy, and the lingering bitterness left them ready for another round that each was determined to win, seemingly at any cost. Outright hatred replaced mere dislike or distrust, and peaceful solutions fell victim to violence against property and individuals. A few more (albeit minor) incidents marred labor/management relations as the century came to a close, further exacerbating tensions between the two groups. Colorado was a tinderbox ready to explode. The union, meanwhile, faced long odds in a struggle that had become largely one-sided, given the state's authority backing the owners.

Still, not all union locals were so contentious and radical. In the northern 1890s Boulder County gold mining district of Eldora, the Western Federation of Miners gained a toehold. The district had more enthusiasm and hustling promoters than high-grade ore, but that did not matter to those caught up in the mini-excitement. The Eldora Miners' Union was organized in the camp on March 8, 1898, "in view of the fearfully hazardous nature of our vocations, premature old age, and many ills, the result of unnatural toil." They formed the association for "the promotion and protection of our common interests."

The local's constitution provided clear insight into what the miners desired from their union, which cost $2.50 to join. Members had to be elected by a majority vote, and were "entitled to a fair trial for any offense involving suspension or expulsion." A meeting was to be held once a week; a $75 "funeral benefit" was established; any member injured or sick would receive $10 per week, "for a period not to exceed ten weeks in any one year." Nothing, however, would be paid for sickness or accident "caused by intemperance or immoral conduct." Officers' duties, rules of order, and other articles filled out the twenty-three pages of the constitution.[5]

In the larger political arena, the fight over silver and gold continued unabated. Coloradans waited anxiously for the 1896 presidential election, when they were sure silver would triumph, and eagerly anticipated the final rout of the gold bugs. The issue was immediately set forth when the Republicans nominated William McKinley, who ran on a gold platform. That left the silver issue up to the Democrats, who met in hot, humid Chicago in July. They were not unified behind the silver banner until Nebraskan William Jennings Bryan rose to speak, in what became one of the transforming moments in American political history. He concluded with the words that quickly enthralled Coloradans and turned into their rallying cry.

> If they dare to come out in the open field and defend the gold standard as a good thing, we will fight them to the uttermost. Having behind us the producing masses of this nation and the world, supported by the commercial interests, the laboring interests, and the toilers everywhere, we will answer their demand for a gold standard by saying to them: You shall not press down upon the brow of labor this crown of thorns; you shall not crucify mankind upon a cross of gold.[6]

For a moment stunned silence greeted Bryan, then a cheering, yelling, boisterous demonstration burst throughout the hall, with the delegates carrying Bryan on their shoulders. The "silver tongued" orator would be nominated on a silver platform, and the issue now rested with the people. Before the convention was over, the Democrats had stolen the Populist reform program as well, and the Populists were reduced to also nominating Bryan, though they stubbornly selected a different running mate.

The fever, the excitement, the expectations of that campaign have seldom been equaled. Poet Vachel Lindsay captured the moment when the debtor West stared angrily at the creditor East and looked to a silver savior.

> I brag and chant of Bryan, Bryan, Bryan
> Candidate for president who sketched a silver Zion,
> The one American poet could sing outdoors . . .
> And all these in their helpless days
> By the dour East oppressed,
> Mean paternalism
> Making their mistakes for them,
> Crucifying half the West,
> Till the whole Atlantic coast
> Seemed a giant spiders' nest . . .
> July, August, suspense.
> Wall Street lost to sense
> August, September, October,
> More suspense,
> And the whole East down like a wind-smashed fence.

When the voters had their say, Bryan received almost 6.5 million votes and carried twenty-two states. Coloradans rallied to the cause with a fervor unmatched before or since, with 83.6 percent of their votes going to Bryan. Many mining counties gave their hero an even higher percentage. It mattered not: McKinley, with more than 7 million votes, carried twenty-three states and a majority of electoral college votes. His strength lay in the high-population states of the Midwest and East.

> Election night at midnight:
> Boy Bryan's defeat.
> Defeat of western silver. . . .
> Defeat of the aspen groves of Colorado valleys,
> The blue bells of the Rockies,
> And blue bonnets of old Texas,
> By the Pittsburgh alleys.[7]

The agrarian and mining states simply were not populous enough to carry the election. Silver would never rise again; it would not be resurrected for another national campaign; Colorado and the other silver states found themselves on the outside looking in at a changing urban-industrial America.

Putting that misfortune behind them, Coloradans could now pay more attention to the mining situation. It was definitely not as gloomy as Coloradans perceived and the silverites proclaimed. Without question, small silver mining districts, such as Animas Forks, Lulu, Vicksburg, Ruby/Irwin, and Montezuma, looked more to the past than toward the future. So, too, did marginal gold districts like Hahn's Peak and Apex. However, offsetting these cases was flourishing Cripple Creek, joined by the revived San Juans. Mining there evolved from gold to silver and then, amazingly, returned to gold, coming in second only to Teller County in gold production. As the *Engineering and Mining Journal* forecast in its July 12, 1890, issue, "San Juan mines have never before looked as promising as at the present time." The shift from silver back to gold was taking place already, especially in the rich triangle with Ouray, Silverton, and Telluride at its points.

In Telluride, three mines in particular rose to national fame: the Tomboy, Smuggler-Union, and the Liberty Bell. Located high in the mountains, all at or over 11,000 feet and liable to be cut off by winter storms, they were so rich that work went on without letup regardless of conditions. Following the general trend, each mine was owned by a company controlled by outsiders, not Coloradans.

Even Telluride's small neighbor, Ophir, prospered. New properties were opened, and older ones saw "renewed activity," reported the *San Miguel Examiner* on Christmas Day, 1897. With typical optimism, the article concluded that "prosperity and growth are now but a matter of two to three years [away]." Even older districts such as Summitville and Carson attracted a few optimistic articles.

Besides Telluride, though, only Ouray and Silverton had any real reason to brag. Thomas Walsh's Camp Bird was becoming one of the great mines in San Juan and Colorado history, and at Silverton, the Silver Lake was only a step or two behind. New mills were built in the nearby valleys, often connected to the mines by long trams.

These aerial tramways became a normal feature of Colorado mining in the 1890s. Sporting ore buckets running along cables, strung between towers, they climbed and circled mountains and spanned canyons while connecting mine portals with mills, smelters, and railroad sidings. Despite being vulnerable to snow slides, they allowed ore, supplies, and even miners to be transported quickly and easily. Well-known mining engineer and reporter T. A. Rickard said of them, "These numerous aerial ropes spanning the intermountain spaces like great spiders' webs, are an important feature of mining in the San Juan region." Indeed, the region has been called "the tramway capital of the American west" because of the large number of trams built there.[8]

With these mines so high and isolated from civilization, the companies had to provide for their workers. Miners at the Tomboy Mine lived in a nearby boarding house. They had access to a YMCA, general store, and bowling alley—all within their building. The Silver Lake Mine, high above Silverton, offered a four-story boardinghouse capable of accommodating 300 men and a dining room that could seat 250. The kitchen was "supplied with every convenience for cooking," and a pipe line furnished "healthy water for domestic purposes." The boarding houses, and "all inhabited buildings," were heated by steam and had hot water for washing. The owners, the Stoiber brothers, did all this in the hopes that "such accommodations" would provide them with a "superior class of miners and mill men" who would stay put and not tramp off to the next rush.

Nothing better illustrated the change in Colorado mining than what was occurring in Cripple Creek and the San Juans. Mines in Boulder, Gilpin, and Clear Creek Counties still mined gold, with Gilpin producing in the $1 million range, but they were no longer newsworthy. Beyond that were declining districts and ghost towns—with one exception. That was Lake County, which made the successful transition to gold mining. From 1894 through 1917, both gold and silver production topped $1 million; for two or three of those years, both surpassed $2 million. Further, Lake County's production of lead, copper, and zinc was the best in the state. Lake County had shown "great advancement" since the troubles of 1893 and 1896.

In these transition years, between hand drills and power drills, the old ways and new ways, a working "stiff," Frank Crampton, left an account of his Cripple Creek days. It gives a very good idea of what it like to work underground.

> Single jacking was a one-man job with no resting, but the double jacking gave some rest with striking and turning being alternated at one or two minute intervals.
>
> It was hard working by candlelight that flickered every time one moved, but there was nothing else used for light. The mine operators issued three candles to a shift, sometimes four at the better mines, often only two at the penny-pinching outfits. If the candles issued burned too fast, one had to work alongside a stiff who had a candle left, or in the dark.
>
> It was harder getting used to the smell of dead power smoke and the reeking, water-soaked timber, but I did.[9]

The "working stiffs" faced death every time they went underground. Though not nearly as dangerous as coal mining, hard-rock mining took a constant toll on miners. For example, the November 16, 1894, *Creede Candle* told of four bodies being recovered eighty-two days after the miners had been crushed by debris that fell when the shaft timbering and shaft house caught fire. "The inexcusable loss of life in the Cripple Creek district is not due to the heartlessness of the mine owners as many believed," claimed the *Cripple Creek Star* (January 1, 1900). Rather, it was "due to their ignorance of mining methods and the incompetence of mining superintendents."

Such arguments, which pitted miners against owners and companies, helped stimulate the union movement in Colorado and elsewhere. Interestingly, when a panel was summoned to review the causes of an accident, invariably it determined that the death or deaths were caused by "careless actions" of the deceased. Usually, the widow and the family received some small compensation, but then were left on their own. Anne Ellis's experience after her husband was killed in Cripple Creek was typical for the time.

> A lawyer from the mine [came] with a paper for me to sign, but I knew enough not to. In these times there was no Compensation Fund. I do sign this paper later, releasing the mine from any fault in the matter, and they give me six hundred dollars; in addition to this each man working in the mine gives me a day's wages. This I take with a feeling of shame, because I know what a day's pay means to some of their families.[10]

In the midst of all the revived gold excitement, few people noticed a new innovation in Colorado mining: the introduction of dredging. The idea was new to the state, though not to other western mining states such as Montana and California. The dredge seemed the ultimate answer to profitable working of low-grade placer deposits. The Blue and Swan Rivers in Summit County had been worked for decades, with less profit each time. Now the dredge owners planned to use "bulk" mining to make them pay once more.

A power-driven chain of small buckets, mounted on an anchored barge or boat, was lowered to cut into the ground beneath the "pond" on which it floated. The chain could work huge amounts of gravel as the buckets brought "pay dirt" to the surface. A washing plant on the boat then treated the gravel to recover the gold. It was hoped that this method could produce a profit with low-grade gravel where other methods had failed. Unfortunately, as it worked its way along the stream, the dredge left behind its "dung," repulsive piles of washed rock that would mark its course for decades.

Breckenridge and Fairplay were the main bases for the dredges, which could operate much of the year except during the coldest winter months when the water levels became too low. Mining engineer Ben Stanley Revett recognized the potential of working the deep gravels along the Blue River. Using Boston capital, he had two dredges built, but they failed because they were too light to handle the gravels.

The dredges potentially offered an economical method to work Colorado's low-grade placer deposits, but they left an environmental mess. That disarray shocked even some turn-of-the-century Coloradans. Up north in Alaska, poet Robert Service described destruction similar to that occurring in Colorado:

> And there a giant gold-ship of the very newest plan
> Was tearing chunks of pay-dirt from the shore.
> It wallowed in its water-bed; it burrowed, heaved and swung;
> It gnawed its way ahead with grunts and sighs;
> Its bill of fare was rock and sand; the tailings were its dung.[11]

Notwithstanding all these innovations, the key to success still lay in reducing the ore to profitable minerals. For decades, the need for more reduction works had been of primary concern. In the decade of the 1890s, some of the most modern plants in the country were located in Denver, Pueblo, Leadville, and Durango; the latter held the title of a regional smelter center, as it was tied by railroads to all the major San Juan mining districts. Many still voiced complaints about reduction charges, freight costs, and other matters, but the smelter shortage crisis appeared to have been resolved.

As often happens, appearances were deceiving. Colorado found itself caught in the middle of the growing American trend of consolidation and the rise of big business—a natural outgrowth of progress, according to some. It was happening in the oil business, the steel industry, railroad lines, and elsewhere. Although consolidation promised better products, more efficiency, and sometimes lower prices, it came at the cost of less competition and the end of the American dream for many. The individual enterprise, small business, and local company had almost no chance against the national monopolies and trusts.

It happened right before the eyes of concerned Colorado miners and mining communities. Smelting had become increasingly complicated and technical, with the introduction of new processes such as chlorination and cyanidation, and a host of new machines working a variety of ores. It had become a big business in the state and throughout the West, and what happened in Colorado paralleled what was occurring elsewhere.

Prosperous smelting corporations moved to buy out or drive out less successful rivals in an effort to achieve a larger and more efficient organization. Suddenly, in April 1899, Coloradans learned abruptly what big business and monopolistic control would mean to them. The American Smelting and Refining Company, incorporated in New Jersey (joining numerous mining companies based in that state, which had few regulations), gained control of smelters in Denver, Pueblo, and Durango. Overnight, it became the nation's largest smelter corporation. Among the major smelters, only the Guggenheims and their operations were not subsumed into the AS&RC. Competition among the smelters virtually disappeared, as did the choices as to where miners and mining companies could take their ore. More than ever before, corporations dominated all aspects of Colorado mining. Complaints arose immediately, but little could be done.

Another developing trend, the abandonment of railroad lines, did not bode well for the industry either. For years, some of the more marginal routes, built to tap a district on the assumption that development would occur, had been running in the red as that promised development failed to transpire. With steadily decreasing passenger and freight traffic, these lines proved a burden on the companies. Railroad companies, like smelters, were being consolidated under ownership that would not maintain unprofitable business that had little future.

When a district lost its railroad, the cost of living, mining, and everything else went up, and it became harder to convince investors that the area had a prosperous future. Isolation once again stared residents in the face. Locals watched as first fewer trains arrived, then trains became combination of freight and passenger cars, and finally the depots closed, with the last train disappearing down the canyon or valley leaving only a whistle's echo and a smoky tail that soon drifted away like the era it was ending.

One piece of "modern" technology remained. With some initial reservations, Colorado mining had adopted electricity as the answer to the old problem of an economical power and light source. Managing the Gold King Mine high in the mountains southwest of Telluride, Lucien Nunn tired of losing money because of the expense of hauling fuel. He tried a different approach: After discussions with the Westinghouse Electric Company and some experimentation, Nunn built a hydraulically generated electric plant in 1891 at the little San Juan camp of Ames. He used alternating current rather than the more popular direct current, because it could transmit higher voltages more cheaply and easily over longer distances.

Soon Telluride was lit, as was Aspen, which was already using lights based on direct current. In the years thereafter, electricity proved a godsend to mining and to communities as well, although it took trial and error to make it work successfully and safely. Bare wires strung through wet mines gave miners quite a jolt (sometimes killing them) if they stumbled into one, or touched a hot wire.

Some of these safety concerns were dealt with when Colorado's legislature took the positive step, in March 1895, of creating the state Bureau of Mines and the Commissioner of Mines to replace the weak "inspector of metaliferous mines" post established six years earlier. The new commissioner was charged with supervision of mine inspection and enforcement of mine safety and health laws. His jurisdiction included the mills, mines, smelters, sampling works, rock quarries, and railroad tunnels of the hardrock mining industry; it did not include coal mining. The Bureau was also charged with collecting and exhibiting mineral specimens and mining data. Writing and publishing books were on the commissioner's agenda as well. These projects, however, were often shunted aside in the effort to accomplish the main objective of supervision.

By the close of the 1890s, Colorado had experienced its greatest gold and silver mining decade. When 1900 was added, it became the greatest period in Colorado's history. The three best years were 1898, with more than $37 million in production; 1899, which topped $40 million; and 1900, which set the all-time record of more than $41 million. Even in their wildest dreams, the fifty-niners could not have imagined this bonanza.

With typical optimism, many Coloradans assumed that the best days were yet to come, even though Colorado had a rival far to the north. They need not have fretted about the Klondike rush, to this "unknown artic region," taking away investors and miners. As Cripple Creek's *Morning Times* told readers in its Thursday, December 16, 1897, edition: "Why not divert as much of its [money] as possible to Cripple where results are certain. It will only require the magic touch of money to start dozens of mines here on a career of production that will astonish the world, and so swell the output of the camp."

Cripple Creek was indeed magical in those days, but its time on center stage was nearly done. Nevertheless, Colorado mining had come a long way in the decade that ushered in the new century. A headline in the *Denver Republican* (January 4, 1900) proclaimed, with obvious pride, "Colorado Mineral Output Greatest in its History." The article opened with eager anticipation and enthusiasm, gushing that "the mining interests of the entire state have never been in better condition." Colorado's greatest gold district and flourishing mining town proudly concurred. The *Cripple Creek Star* boldly predicted, on January 2, 1900: "Paste this in your hat for reference; Cripple Creek will have the biggest boom in its entire history during the summer of 1900."

"History is the witness that testifies to the passing of time; it illumines reality, vitalizes memory, provides guidance in daily life, and brings us tidings of antiquity."
—CICERO (ROMAN STATESMAN, ORATOR, AND PHILOSOPHER)

Photographic Essay
Nineteenth-Century Colorado Mining

The following photographs provide all that Cicero hoped, except, perhaps, "guidance in daily life." Historians and Coloradans are fortunate that the development of Colorado mining paralleled the development and improvement of photography.

Photographers managed, it seemed, to be everywhere to record Colorado mining at its nineteenth-century acme. The result is an amazing heritage of a vanished day and time, of places that no longer exist, of mines that poured forth millions or cost their owners more money than they ever took out of the ground, of dreams that died and of dreams that lived. The result is a fascinating look at a long-departed era, in which the subjects remain ever young.

The Gregory Lode, where it all started. By 1861, many small mines and claims crowded the lode. The large building at the lower left is the Gregory Store and Bank.

COURTESY OF DENVER PUBLIC LIBRARY, WESTERN HISTORY COLLECTION

Sluices crisscross this Gilpin County ravine, and an unidentified small mining camp clings to the hillside in the background of this 1860s-vintage photograph.

COURTESTY OF DENVER PUBLIC LIBRARY, WESTERN HISTORY DEPARTMENT

The ore sorting room at the Caribou Mine. Young boys started work here, often joined by injured miners who could no longer work underground.

COURTESY OF DUANE A. SMITH

Nathaniel Hill brought Colorado mining up to date at his Black Hawk smelter.

COURTESY OF THE PETTEM/RAINES COLLECTION

Burros and mules hauled freight to isolated mines high in the mountains and brought ore out; they were continuously used well into the twentieth century.

COURTESY OF COLORADO MINING BUREAU

The arrival of the railroad promised better days for local mining. The "highline" between Durango and Silverton was a major construction project for the Denver & Rio Grande.

COURTESY OF LA PLATA COUNTY HISTORICAL SOCIETY

Hydraulic mining allowed working of low-grade deposits, but left an environmental mess for later generations.

COURTESY OF MARK AND KAREN VENDL

Most Colorado mines did not "pan" out a bonanza. This is the Argentine Mine at Leadville.

COURTESY OF MARK AND KAREN VENDL

The hoist operator was one of the most important individuals at any mine. Miners depended on him to lower them safely and to bring ore to the surface.

COURTESY OF DENVER PUBLIC LIBRARY, WESTERN HISTORY DEPARTMENT

Miners on the cage, waiting to be lowered to the working level for the start of their shift. This crew worked at Central City's Saratoga Mine in 1889.

COURTESY OF DENVER PUBLIC LIBRARY, WESTERN HISTORY DEPARTMENT

Kokomo, in Summit County, was one of many small camps that dotted the Colorado mountains. Most of these camps had a short lifespan and few profitably producing mines.

COURTESY OF DUANE A. SMITH

The burro could be used for many things besides hauling freight and ore, as this Ouray scene clearly displays.

COURTESY OF OURAY COUNTY HISTORICAL SOCIETY

Cripple Creek's Battle Mountain and the famous Independence Mine. Winfield Scott Stratton's home is in the foreground.

COURTESY OF MARK AND KAREN VENDL

The Strong Mine was blown up during the 1894 Cripple Creek strike.

COURTESY OF CRIPPLE CREEK DISTRICT MUSEUM

The heart and soul of any mining enterprise: the miners. This group posed at Sneffles in Ouray County.

COURTESY OF COLORADO MINING BUREAU

The stuff of legends: A lone prospector sits with his dog.
COURTESY OF DENVER PUBLIC LIBRARY, WESTERN HISTORY DEPARTMENT

10

1900–1999: Looking Forward into Yesterday

As the new century dawned, only Cripple Creek and the San Juans upheld Colorado's mining reputation of yesteryear in the present. Other districts looked more to their past and relied on history to entice investors, along with their finances, for tomorrow. Still, every spring, doggedly optimistic prospectors and miners re-caught mining fever. Finding an antidote was getting harder, though. No new camps or districts were opening, and much of Colorado had already been exhaustively prospected, examined, probed, and dug. In May of 1911, both the *Denver Republican* and the *Durango Herald* asked the same question: "What has become of the genius prospector?" The articles continued: "Months, years have gone and not a whisper hardly about a new gold field or a mining discovery in many parts of the state. The prospector seems to have disappeared; he is not to be found in the mountains and he fails to appear at the assay office."

They did not have to look far for the answer. Mining was now big business, and men labored for corporations instead of seeking private bonanzas.

The surface and easily accessible deposits had already been found and worked, so more costly, deeper exploration was required to find untapped deposits—usually in districts that had been prospected and mined for several decades or longer. Luring financial backers into older districts was much more difficult than attracting them to exciting, flourishing Leadville or Aspen. Furthermore, as each year passed, mining became more expensive and technical. Equipment, reduction charges, wages, transportation—all costs were rising and, for the majority of mining operations, were incurred in working progressively lower-grade ore.

Mining was also falling under the control of the smelters, much to the dismay of those on the digging side of the industry. The American Smelting and Refining Company (AS&RC), which was organized in 1899, had quickly consolidated as many smelters as it could under its control, in the name of efficiency and cost reduction. The company soon controlled about two-thirds of the industry; a "smelter trust" had been born. The Guggenheims and their smelters were just about the only hold-outs against the new behemoth.

When a strike broke out against ASARCO in 1899, the Guggenheims remained in operation while their rival suffered. The well-financed Guggenheims used their newly enhanced position to good effect in negotiations between the two companies; and in April 1901 they merged, with the Guggenheims taking over management of the entire enterprise. At that point, they controlled the industry from Mexico to Montana, and opponents quickly claimed that it was a monopoly in restraint of trade.[1] It was indeed a monopoly, and it made the Guggenheims, who also owned the Silver Lake Mine in San Juan County, premier players in the Colorado industry.

With its total dominion in Colorado, ASARCO could set prices and other conditions as it wished, much to the disadvantage of local mine owners. Perhaps the most damaging development was the closing of smelters. Eleven had been operating when ASARCO took over. The company promptly closed and dismantled six of the eleven; the remaining facilities, the miners complained, were insufficient to handle daily Colorado mine production. The miners protested in a variety of ways, but there was little they could do. As a result, ASARCO gained the nickname "American Screwing and Raping Company," and was cast as the villain that deprived honest miners of their "just rewards," forced mine closures, and hampered development through its monopoly and "unfair practices."

Even with a monopoly, innovations in ore refining continued. Cripple Creek became one of the first districts to use cyanidation. Both that process and chlorination had been experimented with during the 1890s, in the ongoing quest to find the cheapest and most successful method for saving a high percentage of gold. Cyanide won out, primarily because of its strong affinity for gold and the fact that it was a stable, reasonably cheap compound that was not exceptionally dangerous when highly diluted.

After the turn of the century, cyanide was percolated through the old dumps and tailings piles, to dissolve out nearly all the gold that had been missed earlier. Thereafter it proved relatively easy to separate the gold from the cyanide. Meanwhile, the Tomboy mill was experimenting with a new cyanide process to free the gold; eventually, this would become the solution to profitable working of low-grade deposits. Cyanide mills now appeared at the mines. The crushers and stamps that had been kept were eventually replaced by ball and rod mills, and the pans, tables, and other equipment that had long been part of the milling process were discarded. Big wooden vats and steel cyanide drums became the hallmarks of a mining district. The success of cyanidation created a revolution in milling and set off mini-rushes throughout the Colorado gold districts, both placer and hard-rock. Old districts temporarily revived, albeit with fewer miners and support people.

The public again looked at mining with interest and again proved the truth of the old saying, "There's a sucker born every minute." Mining had always had its share of con men and swindlers, but their numbers seemed to increase when the pitch was most alluring or the district in bonanza. Perhaps the most famous mining fraud in Colorado history was the "great diamond hoax" of 1872 in the unsettled northwestern part of the state. Diamonds and other jewels, both cut and uncut, were scattered about the landscape, then "discovered." Before the scam was exposed, at least twenty-five companies had incorporated to mine the supposed diamond fields. With tongue in cheek, the *San Francisco Chronicle* (December 11, 1872) concluded: "Stock robbers are common; anybody can steal at stocks, [t]o salt a gold mine or rob a bank demands no genius. . . . But to plant diamonds from Golconda and rubies from the Orient in the desert place and make them to blossom[,] to our mind, is the highest evidence of business capacity."

How could the public fall for such an audacious dupe? It was simple, according to con man George Graham Rice himself, who had mastered his trade in Nevada mining districts:

> You are a member of a race of gamblers. The instinct to speculate dominates you.

> You feel that you simply must take a chance. You can't win, yet you are going to speculate and to continue to speculate—and to lose. Lotteries, faro, roulette, and horse race betting being illegal, you play the stock game. In the stock game the cards (quotations or market fluctuations) are shuffled and riffled and stacked behind your back, after the dealer (the manipulator) knows on what side you have placed your bet, and you haven't got a chance.
>
> Modern get-rich-quick finance is insidious and unfrenzied. It is practiced by the highest, and you are probably one of its easy victims.[2]

Investors should have paid attention to a 1907 book, *The Economics of Mining*, which offered advice from many specialists, including T. A. Rickard and Herbert Hoover. Among other pearls of wisdom, it offered the following cautions:

> Don't "take a flyer" in mining, but invest your money with the same care and discretion you would use in buying bank stocks, real estate or a silk factory.
>
> Don't invest in a mining company that guarantees dividends. Dame Nature has something to say about *that*.
>
> Don't invest money on the strength of a printed prospectus or the advice of an "interested friend," without preliminary investigation by a reliable engineer.

Leadville proved the value of this advice in various ways, as did some of the other major mining districts. A fine line exists, however, between outright fraud or misrepresentation and mere overenthusiasm. Owners might have convinced themselves that their claims were about to become bonanzas, or they may have failed to understand that the surface assay returns did not guarantee that the ore value would stay high as the miners dug deeper into the earth.

Cripple Creek ore figured prominently in some notorious swindles, one of which used it to promote a Missouri "mine." Walter Scott, better known as "Death Valley Scotty," also used this truly rich ore in spinning the tale of his "secret" desert vein of "fabulous" gold values.[3]

Mine "salting" happened far too often. This could be done in various ways, such as enriching samples with coin fillings or some other materials, planting proven ore from another mine, hiring an unethical mining engineer, or relying on old-timers' "expert reports." Fraudulent assay reports were not uncommon, either; some assayers could find "marvelous" amounts of gold and silver in any rock submitted to them. Offering a mine with plentiful ore reserves for sale and then gutting it before the sale was consummated was yet another way to separate unwary investors from their money. Only the perpetrator's imagination limited the possibilities.

Stock watering and manipulation also occurred all too frequently, in both large and small districts. With an infamous history of bubbles and scams, Colorado mining stock exchanges were perceived by many as no better than casinos. A Colorado judge agreed, and in 1901 he shut down the Denver Mining Stock Exchange by enforcing an anti-gambling law! Insider trading, as happened at Leadville, depressed stock values and left those not in on the plan holding stock that was worthless or nearly so. Insiders fleeced outsiders and even other insiders. As T. A. Rickard wrote in the *Mining and Scientific Press* (December 24, 1898): "A man usually buys a mine not because it is worth the price he gives for it, but because he is justified in the expectation of finding someone who will pay more for it." He concluded: "So the game proceeds. When the sequence has been exhausted, some one gets badly bitten." As with most such ventures, greed was the force driving both the scammers and the scammed.

By the 1900s, the schemes were focused more on getting money for development or from stockholders who did not want to lose their investment. In the 1920s, when many Americans were spending beyond their means and investing heavily in the stock market, fast-talking mine speculators had a field day. The cons might have been small-time affairs compared to earlier frauds, but the victims did not feel or fare any better.

In the end, Rice was right. Greed, naïveté, slick-talking owners and stockbrokers, along with the fact that most mines and mining properties never developed into profitable ventures, led the gullible, voracious, and incautious to make investments in ventures far beyond their level of knowledge or experience.

These scams were not limited solely to mines. Throughout Colorado and the rest of the mining West, promoters of "wonderful, miraculous, amazing, economical" new smelting methods attracted—and fleeced—unwary and sometimes desperate mine owners. Numerous buildings and smelting relics littered the Colorado landscape, monuments to some process that had failed to work.

Meanwhile, underground miners risked their lives working in both profitable and unprofitable mines. They probably knew better than some of the owners what the future held for the property, but the most important matter to them was their pay. In the flush opening days of a district, when there might be a shortage of skilled miners, the pay was sometimes higher than the typical late-nineteenth-century wages of $3 a day (or $3.50 in a wet mine). In isolated mines or those tucked high in the mountains, where miners roomed and boarded on company property, a dollar of their daily wages went for living expenses.

In 1902, the Colorado Bureau of Labor Statistics published a report on wages. Miners received $2.50 to $4 a day, with an "average of about" $3 per day. Machine men were paid $4, trammers $3, blacksmiths $4, timbermen $3.50, and engineers $4. These wages were for eight hours, except for the blacksmiths, who worked nine, and the engineers, who put in twelve-hour days.

Despite the appearance of unions and a slowly rising cost of living, miners' wages improved only slightly from the turn of the century into the 1920s. Reports to the Colorado Bureau of Mines in the 1920s give a picture of the wage scale. For a day, or eight-hour shift, timbermen received $4.50 to $5, trammers and muckers $4, shift bosses $5 to $6, miners $4.50 to $5, blacksmiths $4.50 to $5, superintendents $10, and millmen $3.50 to $4. Some miners chose to receive a percentage of the value of what was mined monthly; others agreed to a contract for a specific job or time period rather than daily wages. They always had to watch the company, though, because some closed quickly and left their workers in the lurch. The miners could always resort to legal action, but, as the saying goes, "you can't get blood out of a turnip."

Accidents continued to take a toll on underground and surface workers alike. Falling rocks, machine failures, unexploded dynamite, carelessness with blasting caps, hoisting-cage problems, company negligence regarding working conditions, and general rashness and lack of care all caused mishaps. In *Tomboy Bride,* Harriet Backus described her husband's work in the Tomboy's lead and zinc mill. George developed lead poisoning, "and for several weeks suffered excruciating pain and was unable to work."

Many miners coughed away their days, victims of silicosis from filling their lungs with rock dust. Silicosis led to other health problems and, very frequently, to death. The problem became particularly acute with the introduction of power drills that literally surrounded the miners with clouds of dust. Eventually, to reduce the dust, drills were designed with hollow cores that had water running through them, producing a muddy mixture as the drill dug into the rock face—but the innovation came far too late for countless old-time Colorado miners.

Although safety and health issues increasingly drew attention, the labor/management struggles of the 1890s, at Cripple Creek, Leadville, and smaller districts such as Lake City, had not solved the basic issue of who would dominate. Once management became organized, the tide turned in its favor, which made outspoken union leadership more determined than ever. Feelings ran deep on both sides. Militant unionists fought entrenched management—first with words, then with strikes, fists, guns, and dynamite. Management responded to activities of the "un-American" unions with company guards, spies, the National Guard, blacklisting, conservative courts, the generally pro-management Colorado government, and violation of their opponents' civil rights.

Between 1901 and 1904, labor struggles cost the state of Colorado a conservatively estimated $750,000. The state acquired a reactionary, violent reputation, and its major mining districts suffered severe setbacks from which some mines and individuals never recovered. Death, destruction, and damage followed in the wake of these contests. It was the worst of times for relations between miners and owners.

In May of 1901, the Western Federation of Miners (WFM), which had gained a stronghold in the San Juans—with locals at Ophir, Rico, Ouray, Durango, Silverton, and Telluride, for example—called out its members at the Smuggler-Union Mine above Telluride. Management insisted that the contract, or fathom, system be adopted instead of daily wages. The union responded that under this system, miners could not earn even the $3 daily wage current in the district. Further, in the hurry to mine a fathom (6 feet by 6 feet and as wide as the vein), the possibility of an accident increased dramatically.

Unlike previous strikes, this one ostensibly had nothing to do with raising or lowering wages, but instead focused on mining methods. Union leadership offered to submit the dispute to arbitration, which management stubbornly and shortsightedly refused. Thus at an impasse, the strike entered its second month.

The Smuggler-Union's manager, the experienced, anti-union Arthur Collins, then hired nonunion miners at $3 for an eight-hour day, something he had refused to grant union members. The WFM countered by trying to induce the "scabs" to quit. That tactic failed, so on July 3, approximately 250 miners and supporters organized to confront the night shift as it came off work.

John Barthell, a striking union miner, stood up and shouted for the scabs to come out. Company guards immediately opened fire, killing Barthell and igniting a shoot-out. When the smoke cleared, two more were dead, six wounded, and eighty-eight nonunion men had been captured, beaten, and brutally driven from the district. Each side blamed the other—small consolation for the dead and injured. There was no traditional July Fourth celebration that year in Telluride, nor did Governor James Orman accede to the owners' demand that the Colorado National Guard be sent in; instead, he sent an investigating team to Telluride to hold hearings. The result was a conference at which the $3 wages and eight-hour day were conceded for underground miners, but not for surface workers and millworkers.

The strike was called off, but both sides remained bitter, edgy, and angry. Telluride's local emerged as one of the strongest in Colorado and served notice that it behooved miners to join. The owners responded to this threat by organizing the San Juan Mining Association covering San Juan, Ouray, and San Miguel Counties.

Telluride became a hotbed of contention as both sides jockeyed for dominance. The union boycotted the vehemently anti-labor *Daily Journal* and threatened any business that would not display a "fair card" in its window and join with a union boycott. A Business Men's Association responded by supporting the paper and pro-management businesses.

Then, on November 19, 1902, Collins was assassinated at his home. With no proof whatsoever, the owners blamed the union and had its leadership arrested, while the real murderer—unknown to this day—escaped. Of course, the union leadership denied any involvement in the murder. The accusations and counter-accusations growing out of Collins's murder further inflamed an already volatile situation. The line had been crossed: The union would either be destroyed or would control the district—and the owners were not about to let the latter happen.

Collins was replaced by the flamboyant, controversial Bulkeley Wells—a ruthless anti-union man, much admired and much hated. Both sides braced for a final round, which came in September 1903 when the WFM attempted to gain an eight-hour day for millmen. The managers stood firm, preferring to keep their plants closed rather than "submit to the [union's] dictates." Wells confidently predicted, "The strike here and elsewhere . . . promises to be an absolute failure."

When the union at Cripple Creek went out on strike concurrently with Telluride, the labor/management struggle had turned the state's two leading districts into battlegrounds. At Telluride, the prolonged strike spread to nearby Ophir, while armed guards and armed union pickets stared each other down. Among those who arrived was Robert Meldrum, a professional killer imported by the mine owners, who reportedly made his local debut at the Cosmopolitan Saloon by ordering a drink and announcing to startled patrons: "I'm Bob Meldrum. You can always find me when you want me.

Now, if any son of a bitch has anything to say, spit it out; otherwise, I'm going to take a drink—and alone." According to local legend, the room promptly cleared, with people leaping out windows and dashing through the swinging front door.

The WFM countered by hiring a gunfighter of its own, Joe Corey. With the owners in control of San Miguel County, the county sheriff promptly deputized Meldrum and several of his fellow guards, giving them the considerable advantage of delegated legality over the strikers and their supporters.

As positions hardened, tension mounted, threats turned to violence, and in the owners' minds anarchy was imminent. The call went out for the National Guard. This time, ardently anti-union James Peabody sat in the governor's chair, and he did not hesitate. In marched six companies of infantry, supported by two cavalry companies. They deployed at Telluride, Ames, Pandora, and at various individual mines. These troops unequivocally represented the owners against the strikers. To ease the financial burden on the cash-strapped state, overjoyed owners furnished meals and quarters for the troops until permanent camps could be established.

The change in the power balance soon became obvious to everyone. The owners hired scabs to replace strikers, and striking miners, "mostly foreigners, were arrested and charged with vagrancy and told to go to work, to jail, or to leave."[4] Gambling houses and saloons were boarded up and on the surface Telluride became a ghost town. Civil rights went out the window; people were guilty until proven nonunion in membership and loyalty. Union leaders were seized for threatening nonunion miners. By the end of December, all seemed quiet—but underneath, the community seethed.

On January 3, 1904, Peabody, who had thus far resisted declaring martial law, suddenly did so, claiming that San Miguel County was bordering on insurrection and rebellion. The owners had been pushing for this move, and Peabody, convinced by hearsay and "threats," took action. So did the owners, who organized their own private troops, mustered them in, and volunteered them for state service. This amazing turn of events resulted in a complete victory for the owners. Striking miners were deported, soldiers patrolled the streets, passes were required to be out and about, guns were confiscated, and Bulkeley Wells assumed command of the district. All this was accompanied by selectively enforced press censorship.

The victorious owners, now certain of their absolute dominance, continued to deport union miners. Finally, with only a few remaining, martial law was lifted on March 11. After the militia left, however, not twenty-four hours passed before Wells organized a "citizens' alliance." When some of the deported men returned, the alliance promptly put them on a train, where they were joined by about sixty-five of their supporters. All were ordered not to return.

Grateful friends honored Wells with a victory banquet on March 17. It proved a bit premature, however. When armed men promised to aid returning strikers, Peabody again sent in the guard: Troop A, Wells's handpicked Telluride supporters. They "cleaned up" remaining union members and backers, stopping returning strikers, jailing a few, and concluding a general roundup of "undesirable men in the district." The second reign of martial law was ended on June 15, giving final and total victory to the owners and their backers—but nobody had really won. The spirit that distinguishes an exuberant, prospering community from a stressed and worried one died that year in Telluride.

Colorado's adjutant general, the pompous Sherman Bell, proudly boasted that peace and good order had been fully restored. That was true only in the sense that mills and mines returned to full operation after their owners seized absolute control. The decimated Western Federation of Miners finally conceded in November, calling off the strike. Management did hand the miners a pyrrhic victory when it unilaterally granted employees an eight-hour day, but the concession came at terrible cost: human trauma, monetary losses, violation of civil liberties, brutal repression, and severe damage to the district's reputation.

Bell, meanwhile, played a much larger role in Colorado's more newsworthy dispute: the strike at Cripple Creek, a district still enjoying more than $10 million annually in gold production, a thriving economy, and national fascination with its goings-on. For the miners, the old days, when they could tramp on to another camp or a new district and make their individual fortunes, were but a memory. Corporations now ruled, and that was as clear at Cripple Creek as in Telluride.

Cripple Creek was such a big player on the mining scene that, like Leadville before it, it had its own mining stock exchange. Cripple Creek's most famous millionaire, Winfield Scott Stratton, had built the Mining Exchange Building in Colorado Springs, where the association kept its offices. The 1907 report listed thirty-one pages of companies and promised its readers that:

> All properties listed upon the Exchange are passed upon by the Listing Committee and the Attorneys for the Exchange. Titles, development, location, future prospects and management are examined. Every stock is registered with a responsible Bank or Trust Company.

The report also listed each company's officers (including their salaries), capitalization, dividends property, shares, plant and machinery, people employed, and taxes paid, and stated whether the firm faced any indebtedness or litigation. Investors must have been reassured by such a complete summary, especially considering the number of crooked mining investment schemes being perpetrated at the time. Of course, no one could tell what the future might bring or what might be hidden in the statistics.[5]

Investors may have rested easy, but in Cripple Creek itself matters were not going quite so well, at least from the owners' viewpoint. In a district that was literally a union camp, almost all labor sectors—including the porters and dance-hall girls—were organized into various unions. The upset owners, who well remembered their 1894 embarrassment, were eager to break the unions' growing strength.

The trouble did not originate with the miners; instead, it was the millworkers who stirred the pot. The mill owners had stubbornly resisted all efforts to organize workers at the reduction works for Cripple Creek, which were located at Colorado City rather than in the gold fields themselves. It was there that the WFM eagerly moved to organize the workers. In March 1903, the workers struck against two mills that operated closed shops, and in came the militia. The pattern followed was strikingly similar to what was transpiring at Telluride.

A well-organized, equally determined Mine Owners' Association, along with Colorado's unwavering anti-union governor, Peabody, stood in adamant opposition to the workers. This time they were well prepared. They controlled many local government offices (though not the sheriff). The companies had a fairly substantial war chest, which would hold out if the strike was not too prolonged, and the mine owners guaranteed that they would underwrite some of the expenses, which made it easier for the nearly bankrupt state to intervene.

The strike went through several phases. Initially, it was about recognition of the union. After some hesitation, the 3,500 Cripple Creek miners marched out in a sympathetic strike against the mines supplying the two closed-shop mills. This forced management to make concessions, and the matter seemed settled in April 1903. Then the millworkers walked out, initially against a mill they claimed had not lived up to the agreement. On August 10, unionized labor in Colorado City and Cripple Creek went on strike again, called out by William "Big Bill" Haywood, the union's radical executive secretary. No one knows how many miners actually supported the strike, as no vote was taken.

The strike nearly paralyzed mining operations and was potentially hazardous for all concerned. The district's prosperity depended solely on uninterrupted gold mining, because there were no other industries to fall back on as a safety net. A prolonged dispute might prove catastrophic to the nearly 50,000 people residing in the district. Nor was the union, despite its strength, in a position to withstand a long siege. Union president Charles Moyer optimistically hoped the owners would help bring the smelter people to their senses before the strike went on too long. That plan failed and the strike dragged on. As August slipped into September, any possibility of compromise almost vanished, and both sides prepared for a long, hard struggle.

Not only owners but merchants as well battled the union. Fearful of carrying customers' debt, the shopkeepers initially pressured the striking miners by demanding cash-only sales (typically, credit was offered until the end-of-the month payday). The union countered by opening its own store. Only "traitors" traded with the other side's merchants. Positions hardened, friendships evaporated, and communities became split.

Violence, terrorism, and a mine fire ended any hope of a peaceful settlement. A petition was rushed to Denver to alert Peabody to the problems, while owners demanded that the sheriff hire extra deputies whom the owners would select and pay. In September, Peabody, the "law and order" governor, after sending an investigating committee to assess the situation, responded predictably. In marched nearly a thousand guardsmen "to protect all persons and property from unlawful interference." Colorado thus had two districts under military control.

What that meant in reality was a pro-owner resolution, under the guise of seeing "public peace and good order . . . preserved upon all occasions." The sheriff and others, including Victor's city council, protested that the situation did not warrant such action, but they were crying unheard in the wilderness. By the end of the month, troops patrolled the principal roads and guarded the mines, and the district was virtually divided into two camps, pro- and anti-union. The WFM faced a grossly uneven fight against the owners, the conservative courts, most local newspapers, the vast majority of Colorado public opinion, and the power of the state government.[6]

The *Cripple Creek Times* (September 8, 1903) reported that the Mine Owners' Association blamed the strike on a "few irresponsible agitators" whose methods were "inimical to the rights and obligations" of everyone. Further jobs in the district, it stated uncompromisingly, would be given to nonunion men only. Both sides used such propaganda. The owners and their backers claimed that the "militant" union and its leaders were lawless and un-American. The union fired back with salvos about tyranny, the workers' plight, and the rights of Americans. Meanwhile, the union's resources dwindled as it tried to keep two major strikes going. The owners had a huge advantage in what became a war of attrition.

During this largely one-sided struggle, a campaign of harassment against the strikers gained momentum, with Adjutant General Sherman Bell personally supervising operations. In the name of "military necessity," local government was ignored, the pro-union *Victor Record* was eventually "captured" and destroyed, and civil courts were superseded by military courts. Civil rights were trampled again, pro-union businesses boycotted, and union members blacklisted. *Vagrancy* was so broadly defined that the law could be used against striking miners to force them either back to work or out of the district.

By February 1904, all the pro-management efforts appeared to have succeeded. The few remaining troops were placed under the authority of local officials, and were completely withdrawn in April. Then, on June 6, the uneasy calm that had settled over the district was shattered by the desperate union. Violence erupted anew when Harry Orchard, a professional terrorist, dynamited the Independence railroad depot just as the night shift left work. Thirteen men died, and many other nonunion miners were injured. The *Cripple Creek Times* that day accused the WFM of "cold blooded murder," and the public generally agreed. This ghastly event brought back the guard, martial law, wholesale arrests, deportation, the wrecking of the Victor union hall, and mob harassment of union members. Meanwhile, Bell took extraordinary measures to destroy what was left of the union.

On July 27, troops were finally withdrawn again. This led to one more violent outbreak, this time by the owners' henchmen, in what the *New York Times* (August 7) called a "reign of terror." A union store was ransacked, mobs seized prisoners and abused them with impunity, union sympathizers were terrorized, and Peabody refused to intervene. What remained of the union disappeared, leaving the owners in complete control.

These events spelled the end of the Western Federation of Miners in Cripple Creek and throughout the state. To the public, press, and government, it seemed that everything the owners claimed was true: The union was violent, un-American, radical, and opposed to law and order and the constitution. More importantly, the labor disputes were costing taxpayers and the state hundreds of thousands of dollars.

The strikes at Telluride and Cripple Creek cost the WFM everything it had previously gained. Its outspoken leaders had pushed too hard and too fast, converting strikes for the improvement of miners' wages and working conditions into brutal struggles for recognition and power. The owners' vicious responses, along with exceedingly questionable actions by Peabody, Bell, and Wells, won the war. In the end, though, there were only losers, including the state, state government, and the public's perception of this western commonwealth.

Colorado now had a reactionary image, and unionism had suffered a serious setback, but the miners lost the most. Not only did they have a radical, un-American reputation, having been made villains in the public eye, but they were also firmly trapped under the thumb of management. They were not even free to work: They had to have cards to show they were not in the union. This requirement persisted long after the strikes; in 1912, membership in the Miners' Protective Association was still mandatory in Cripple Creek. Information on each miner included his age, marital status, length of time working in the district, and birthplace. The owners? They suffered the least, except for lost profits and an image problem—which probably did not bother them in the least.

Despite the violence that came and went, in declining or less prosperous districts mining continued much as it had for decades. In the still-prospering districts, though, mining changed more than it had in years. Electricity powered lights, drills, pumps, engines, and cages that moved up and down shafts; it warmed doghouses and propelled ore trains. Trucks supplanted mule and burro trains and even the once-omnipotent railroads. Where the roads were good, it was easier and more convenient to transport ore to mills by truck than by rail. In the next few years, carbide lamps took the place of candles, and the once-standard cloth cap, hardened by pitch or some other substance to provide a little protection, was replaced by the hard hat. Some mines had trained rescue crews in case of an emergency. Nevertheless, even with all these improvements, mining still remained a skilled, difficult, and dangerous occupation.

No other mining district in Colorado could match Cripple Creek and the San Juans in production; only the declining Lake County even came close. Much of Colorado mining mirrored Carroll Coberly's experiences at Ashcroft, nearly a ghost town even then. He went to work in the Montezuma Mine in 1906, and his first job was to clean out a tunnel that had been blocked by ice. It was a "hard miserable job," with the men coming out at night "wet to the waist."

The next summer, a new, inexperienced manager built a mill at the "wrong place," a location where avalanches were common. In October he "got scared" and "closed everything down." By then the company was broke, a new company took over, and Coberly became superintendent. For the next five years, snow and snow slides continued to hamper operations, taking out the boarding house, tram towers, and other buildings. The mine never paid, but Coberly was convinced it would have had the mill been located in a spot "safe from snow slides."[7]

Even as corporate mining began its transition to the modern age, expectant prospectors still took to the mountains, despite the tremendously long odds against finding anything valuable. To help the neophyte, books such as *The ABC of Mining* (1898) included chapters on prospecting, mining, floating a company, and other such necessities. *The ABC* offered tips on camp life, such as: "The working man found out long ago that pork and beans suits him nicely"; "the dictates of fashion [are] unheard [of] on the mountain side, and beneath the pines, dress resolves itself into a mere question of warmth and comfort"; and "a camp kit of cooking utensils often begins and ends with a frying-pan and tin kettle."[8]

Such optimism lent a glimmer of hope to Leadville. Its gold belt had kept it in the news early in the century, but silver—even in the paltry range of fifty to sixty cents an ounce—still remained its main claim to fame. Annual production stayed at the $1 million-plus level into the 1920s. Both operator-owned companies and lessees remained active in the district, but there and elsewhere, increased leasing indicated that a district was past its prime. As *The Mining Investor* (September 16, 1907) reported: "The most interesting sort of prospecting is in progress at this time in various sections of the Leadville district, Fryer hill and the downtown section coming in for their share. There is actually more ore in sight today in Leadville, through careful recent development, than at any time in the history of the camp." The article continued: "But a new era is dawning for Fryer Hill. Its development at depth is inevitable. Leadville in other sections has proved rich at depth, and is now producing to the second contact, which has been found at depths ranging from 1,000 to 1,400 feet." Colorado miners had lived on such expectations and guesses since the Pike's Peak rush nearly fifty years earlier.

Leadville was not the only district living on hope. Gunnison County, for instance, had barely caused a ripple, except for Ohio City. There, the Gold Brick District, with its Gold Links Mine, briefly produced a fair amount of gold. Also, Gold Creek near Ohio City boosted the county's gold production to more than $100,000 each year from 1908 through 1912. That had happened only eight times in more than sixty years of mining in the county.

Dorchester, on the Taylor River, contributed a brief boom after 1900, being hailed as the "coming mining camp."[9] The optimism had some basis, as nearby mines produced well into the century's second decade, but thereafter the low price of silver and the cost of transporting ore to Aspen drove the district into a downward spiral.

Most of the excitement in the Gunnison "Gold Belt" occurred in the 1890s, but the Iris Mine managed to hold on into the new century. Part of the treasure was just across the county line in Saguache County. The *Cochetopa Gold Belt* (August 23, 1895) excitedly described it as "a great big free milling proposition," worked by three mines with steam plants and "two stamp mills of ten stamps each." Yet, by 1903, the district was finished, every mine had closed, and the camp verged on ghost-town status.

More representative of what was happening was Sillsville in the southeastern part of the county. In 1903, Sillsville boasted a camp and nearby mines, of which the Maple Leaf was the best. However, the higher-grade surface ore quickly pinched out, and the miners and companies departed as rapidly as they had come. The district peaked, the boom became a bust, and Sillsville disappeared from the newspapers.

Gunnison County's production was extremely uneven in the years from 1900 to 1920. It produced $124,676 in gold in 1912 and $9,500 the next year. By the first year of the "Great War," 1917, the total had dropped to $6,635. Suddenly, Gunnison "revived" from 1919 into the early1920s, with silver production in the $20,000 to $30,000 range and gold jumping to a high of $30,000. At best, though, those total amounts were only a pale reflection of the county's great silver years back in the 1880s.

Clear Creek County completed the transformation it had begun earlier. Gold production in lower Clear Creek surpassed silver in upper Clear Creek starting in 1902, thanks to the gold mines around Idaho Springs. Annual gold production had been in the $400,000–$700,000 range for a decade before it surpassed silver, and it remained in that range for more than another decade. Sadly, even gold was not the savior the locals had hoped for. The *Georgetown Courier* (January 10, 1920) depressingly summarized what had happened: "The mills at Idaho Springs were not active in 1919, for the Argo and Jackson were closed, and the Hudson and Newton were operated only on part capacity."

Georgetown, Silver Plume, and Idaho Springs newspapers, regardless of the year and situation, remained stubbornly optimistic, as these excerpts illustrate: "The Old Terrible mine is once more the scene of great activity." "Several hundred men are at work in the mines and what insures the great growing prosperity of the place is that nearly all the men are actually engaged in producing pay ore." "The Colorado Central Mining company has so far perfected its pans that work from now on will be vigorously pushed, and within the next sixty days the company expects to have in the neighborhood of 100 men employed." "The Georgetown tunnel is within about 200 feet of the Cliff lode which produced handsomely in early days from surface workings." "This great property which has been idle for seven or eight years is about to be reopened."

Not all the reports were so encouraging and boosting of local self-confidence. "The Colorado Central mine, which has been a boon to Georgetown for several years has suspended operations, or rather, operations were forced to be suspend by the employees quitting." "The aerial tramway to Sunset Peak has been sold as junk to Denver parties." "With the exception of the work [done by Mr. Winter,] all the properties lying between the Colorado Central and the Waldorf are dormant."[10] There were fewer of these sobering articles, and their coverage was not as wide, but they told the underlying truth about what was happening in the county.

Another indication of the decline in Clear Creek County appeared in the 1913 *Colorado State Business Directory*. In Georgetown, once the glittering "Silver Queen," few people in town were directly connected to mining. They included employees of the headquarters of seven mining companies and one tunneling company, three mining engineers, and one individual who simply listed "mining." Nobody gave "miner" as an occupation.

In contrast, San Juan mining, which featured gold production, was in full flower. Ouray's Camp Bird and Telluride's Liberty Bell and Smuggler-Union became famous. Telluride bounced back quickly after the strike. The *Engineering and Mining Journal* (August 15, 1908) reported that it was booming, with mines and mills running at "full capacity" and more men employed in the district "than for many years past." The year before, the *Journal* (August 14) had described how Silverton "has really taken on new life." Poor Rico had not, however. George Backus wrote Harriet when he visited the town, "Rico is a deserted, broken down mining camp." He might have added that its production had sunk to well under $100,000.

Meanwhile, lawsuits—the trouble that had hit so many mines over the years—engulfed the San Juans. The Smuggler-Union and Liberty Bell went to court over issues of trespass and ore removal, all of which, once more, involved the infamous apex question. The Smuggler won, but at the cost of large amounts of money and two years of litigation.

There would be one more old-style excitement: Cave Basin northeast of Durango. In 1913–1916, a mini-frenzy mounted; the *Durango Weekly Herald*

(March 26, 1914) said that it "promises to be one of the best in the state." Durango and Bayfield fought to be the "gateway" to the "bonanza," while prospectors, miners, and investors scurried thither. It all came to nothing: The Cave Basin miners never hit enough pay dirt to move their "mines" beyond mere prospects. It was soon pushed out of the papers by the outbreak of World War I in Europe.

President Woodrow Wilson had hoped that the United States could remain neutral, but that proved impossible. As America's entry into the war neared, the San Juans slowed down. Hard-pressed England required its citizens to bring money from foreign investments home. That dried up a major source of operational and investment funds that had kept the district going for the past quarter of a century.

The World War I era marked the end of large-scale gold mining in Colorado for nearly a century. It dropped from more than $21 million in production in 1915 to $12 million in 1917—a trend that continued downward in the years thereafter. Silver mining enjoyed a revival, jumping from $3 million to $7 million annually during the same years—interestingly, a high percentage of that was gained as a byproduct of mining for other minerals. Given the paucity of financial resources, mining in those days was more frequently carried on by smaller companies and crews.

To see the trend, one need only examine the mining inspector's reports from the 1920s in declining Clear Creek County. The Glasgow Mine in 1924 had one man and "at times two men employed"; the Fenton Mine (1928) employed five men; Sussex Lode (1920) was leased; East Butte was leased; High Grade Group employed two men; and the Atlas Tunnel (1926) had two men employed. Production of these properties was described as "none," "no available ore in sight," "all work in line of development," "the vein has not been intersected yet," "some ore available," and "no ore in sight"—not a very encouraging summary, but typical of this county and others in the 1920s.

Where, then, did county production come from? The leased Gold Belt Mine employed nine men underground and on the surface and operated almost the entire year in 1923. Ore averaged eighty-three ounces of silver and a trace of gold to the ton, along with lead. In fact, it was the production of base metals that kept many of the precious-metal mines operating in Clear Creek County and throughout Colorado. These commodities were far less romantic than gold and silver, but crucial to a mine's production and profitability.[11] As the *Georgetown Courier* of January 10, 1920, explained in summarizing 1919 mining in the county, "In Clear Creek County the low price of lead and zinc closed the principal producing mines at Silver Plume in May 1919, and thus cut off the usual large silver contribution from that camp." That held true for every district in the county where mines were still operating.

Old-timers Boulder and Gilpin Counties followed the same general pattern. Gilpin struggled to maintain $50,000 in annual gold and silver production by the mid-1920s. Boulder County, where mining gradually declined as well, actually managed to pull even with its longtime neighbor and rival. Throughout the two counties, where the mining industry had once been dominant, only traces of it remained.

The industry blamed some of its problems on the lack of smelters. By 1927, only the Durango and Leadville smelters remained in operation. Plants in Denver, Salida, and Pueblo were being closed and dismantled, meaning that only counties "immediately contiguous to the two operating plants" could maintain profitable production with low-grade ore. That situation, the *Colorado Annual Report* (1927) of the Colorado Bureau of Mines stated, had a "far more disastrous effect on the mining industry than all the ill effects arising out of the postwar depression."

Gold fever never went completely extinct, though. In 1922, interest suddenly revived in the "old" Hahn's Peak, which had been promising much but producing little since the 1870s. "There is gold in the Hahn's Peak district. Veins of it, lodes of it—and a mother lode," declared the opening paragraph of an article in the *Mountain States Mineral Age*. Despite cautioning that the gold might be under 250 feet of shale, the article forecast that "when it is found it will mean, in all probability, another Cripple Creek" and insisted that it was "not a wildcat mining" region.[12] Unfortunately, the article's fear that the gold "may never be found" proved entirely correct.

Dredging also raised some people's hopes. At one time or another, dredges working rivers and streams operated in Park, Gilpin, Costilla, Routt, and Summit Counties. A little dredge even operated on the Animas River, though it dug deeper into investors' pocketbooks than into the river bottom. Between 1898 and 1942, nine dredges were operated at various times by several companies along the Blue and Swan Rivers. They even clanked into Breckenridge. As their buckets tore up the water bottom, they could be heard for miles, or as Alaskan poet Robert Service described it, "turning round a bend I heard a roar."

> It wallowed in its water-bed; it burrowed, heaved and swung;
> It gnawed its way ahead with grunts and signs;
> Its bill of fare was rock and sand . . .[13]

While "it glares around with fierce electric eyes," the dredge "looked like some great monster in the gloom." "Full fifty buckets crammed its maw," bringing up the gravel.

However, conditions had to be exactly right to sustain profitable dredging operations. In most places, high hopes and expectations were dashed on the rocks of reality—in this case, large boulders in the streams. Rocks of all kinds shot costs beyond estimates nearly everywhere.

Summit County and Breckenridge saw success in 1902 when, in a short five-month season, the American Dredging Company worked profitably on the Swan River, reportedly recovering more than $100,000 in gold. On the Blue River, the Gold Pan Company finally reached bedrock forty feet below the surface, but encountered large boulders that hampered operations. The company also conducted hydraulic operations on its nearly 1,700-acre property. It had a large dumping ground for tailings, but people downstream still encountered muddy waste tailings floating by. The *Engineering and Mining Journal* (December 11, 1909) concluded that hydraulic mining in the Breckenridge district "has never been successful."

By 1909, three dredges were operating in the area. The amount of gold recovered varied considerably from year to year: from $100,000 to $600,000, depending on the length of the season, the water available, and the amount of gold recovered. The dredges moved a lot of gravel per day, leaving the familiar trail behind. The *Engineering and Mining Journal* (February 8, 1913) reported that a dredge could handle 3,000–6,000 cubic yards per day at a cost of five cents per yard. That was the benefit of a dredge working low-grade deposits: high volume at low cost. Thanks to its dredges, Summit County had some of its best years in gold production just before the United States entered World War I in 1917.

Dredges, though, were frustratingly prone to regular breakdowns, and fires burned more than one boat. Some particularly ill-fated boats burned twice. Lawsuits, debts, angry creditors, disgusted stockholders, lack of funds, and failed hopes trailed the dredges after World War I. The Depression years were even worse, and by the 1940s Colorado's dredging era was over.

After digging into the valleys to depths of sixty feet or more, dredges also left an environmental mess along the streams for future generations. The earlier era saw it quite differently, though; as the *Summit County Journal* had stated back on December 9, 1899: "A debt of gratitude we will never be able to pay, [dredging] transformed Swan River from a worthless barren wilderness to scenes of commercial and industrial activity, the like of which was never dreamed of, three years ago." Locals tended to agree with a dredge boat superintendent who was quoted as saying, "Industry is always to be preferred to scenic beauty."

By the time the Roaring Twenties ground to a halt in 1929, Colorado mining had changed radically. Leadville was still a grimy, industrial town, but even it had slipped under the $1 million annual production level. Gilpin County, once number one, barely reached one-twentieth of that total. Georgetown was promoting tourism, as was Idaho Springs. Several mountain communities would have loved to have done the same, including Breckenridge, Ouray, and Aspen, but they remained isolated far from tourist destinations. Railroads carried few tourists anymore; most came by automobile, and Colorado's roads beyond Denver and the foothills were not too good, particularly those over the mountain passes. Only the adventuresome toured there; the rest stayed on familiar and better maintained roads.

Where miners had once lived, worked, and played, there were only ghost towns, abandoned mines, and a few diehard old-timers, trapped in yesterday. Silver Cliff, Animas Forks, Ruby/Irwin, Vicksburg, Buckskin Joe, Parrott City, and the appropriately christened Sunset—names that had once sparkled in silver and gold dreams—were mere shadows of their former glory.

Tourists already wandered around sites with rotted boardwalks, gaped at mill skeletons, and scrambled over dumps looking for a high-grade souvenir. They peered into mine portals and yawning holes, where once the earth had poured forth gold and silver, and opened doors to abandoned cabins, speculating as to what had happened there and when. They took home souvenirs, then forgot where they had found them. Meanwhile, the remains of the vibrant communities of only a generation ago were disappearing.

A haunting, poignant image of Kokomo in 1912 captured clearly what was happening. Older people were slouched along the plank walks "with a far-away look in their eyes as if they had lost something which was never to come back to them." That look could have been seen at Caribou, Platoro, Gothic, Independence, or any of the dozens of once-promising and prosperous communities that dotted the Colorado Rockies. The description of Tabor's Matchless Mine that appeared in *The Mining Investor,* back on September 16, 1907, serves as an epitaph for an era now departed into history and legend: "[When] the rich ore was exhausted, operations practically ceased, and the ancient dumps, formed thirty years ago, and dilapidated shaft houses are in evidence on every hand as relics of a prodigiously rich period."

11

Mucking through Depression, War, and New Ideas

Mucking is defined as "the operation of loading broken rock [ore, debris] by hand or machine[,] usually in shafts or tunnels." In the 1930s through the 1960s, mucking pretty much described what the mining industry was doing on various levels, from the office, to mining methods and innovations, to miners working at the breast of the drift. Eventually, the industry had to confront the new attitudes of the general public toward mining.

Long before all that happened, however, mining, Colorado, the rest of the United States of America, and the whole world had to face the crash of October 1929 and the subsequent grinding Great Depression, which in many parts of the state lasted for nearly a decade. The industry would eventually get some help from the sweeping New Deal programs advanced by President Franklin Roosevelt, and gold mining would actually make a comeback, but that was in the future. The present reality, in 1929 and 1930, was grim.

Still staggering from hard times during the 1920s, mining, like agriculture, was further damaged when a concatenation of factors caused the

October crash of the stock market in 1929. Those factors pushed the nation over the edge both emotionally and financially. It did not happen all at once. Some people, businesses, and industries were affected in 1929, others in 1930, and the whole state within a year. By then, the economic collapse and the futile efforts to overcome it permeated almost every aspect of life.

Coloradans might argue which was worse, the hard times of the 1890s or the 1930s, but the answer was more or less academic. With people thrown out of work, money lost in bank failures, and homes, farms, and other property foreclosed on and sold under the sheriff's auction hammer, Americans saw only tribulations and troubles wherever they turned. From bread and soup lines, and unemployment running between 25 and 30 percent (or higher), and with a general feeling of discouragement and hopelessness, Coloradans watched dark clouds settle over their land, dreams, and hopes.

The Colorado Commissioner of Mines, John Joyce, thought this the perfect time to rejuvenate state precious-metal mining. While he found it "most encouraging" that old mines were resuming operations, there seemed to be a negative attitude toward the industry among the "business, industrial and financial interests throughout the state." What is needed, he proposed, "is intelligent and loyal support in a wholesome manner, instead of the disheartening way that generally marked their attitude toward mining in Colorado during the past decade."[1] Joyce did not elaborate on why, but the conservative state government and the dominant Republican party of the 1920s had not offered much support to the industry. Nor was the public much interested in mining anymore—with a few individual exceptions.

During some of the darkest Depression days, a young man went to work at the Camp Bird Mine in 1932 to "get a stake" to get married, as he expressed it. In *One Man's West*, Coloradan and author/historian David Lavender captured the lure and the pride of mining as few others have been able to do. A few excerpts yield insights into the miner and his job that could have been noted at any time since the first miner dug into the Colorado Rockies.

> The miner goes to his wet, lonesome, sunless trade with his head up; he calls himself a quartz man, a hard-rock stiff, and considers himself superior to the grubs who toil in softer, easier mediums. . . .
>
> Nothing can convey the impression of the overwhelming darkness. It was not just the absence of sunlight, for the sun had never touched this spot. The top of a mountain, the middle of a desert have their stars, wind, dawn, their feel of space. Here was nothingness. Eternity passes our comprehension. . . .

The mouth of the shaft was surrounded by a wooden floor. Water dripped on this and mixed with the mud to form a slime. Footing was uncertain to say the least, and working around the edges of that three-hundred-foot grave kept me in a dither that no amount of familiarity could alleviate. Fear is a wonderful skin preserver. In certain respects a greenhorn is safer in a mine than a veteran, for the tenderfoot is imaginative. He sees danger in shadows, hears it in every creak of a timber, every thundering explosion. The veteran—pooh, he knows all about those things and they don't bother him. . . . [P]erhaps they should. It is a fact that most mine accidents happen to the oldest hands.[2]

Lavender's experience in the Camp Bird reflected that of the industry. Mining too had no margin of error in those days. No safety net, no surplus, helped it through the hard times. With little or no state or federal assistance during the Republican-dominated 1920s and into the 1930s, the industry stagnated, keeping miners out of work. Their only hope seemed to be the untested, basically unexplained New Deal that Franklin Roosevelt and the Democrats promised during the 1932 campaign. When they won, Washington rode to the rescue, starting with the famous hundred days that saw so many programs instituted. Eventually, the New Deal brought more government involvement in people's lives and in the American economy, including mining.

This became plain to mining when, on March 19, 1933, an executive order prohibited the free export of gold, except for business requirements. That was followed by Roosevelt's ordering all persons to turn over their gold to banks, thus abolishing individuals' right to possess or export gold. Congress took this a step further in June by taking the country off the gold standard, and outlawing clauses in private and public contracts that called for payment in gold.

Mining got some of the help it needed when the price of gold was raised to $35 an ounce on January 31, 1934. That spurred excited interest during these hard times in both placer and hard-rock mining. Even finding a small portion of an ounce promised better times than living off the dole.

Silver also received assistance. When the idea of putting more silver in circulation was proposed as another method of inflating the currency and raising prices, it was almost as if the Populists had been resurrected. Nevada Senator Key Pitman led the charge. With the president having previously been given authority to increase the nation's currency at his discretion, and to coin silver at any ratio to gold he might choose, Roosevelt had the Treasury begin to buy "newly mined domestic silver" on the basis of a 16-to-1 ratio.

In a real sense, Bryan and free silver had finally triumphed. Silver was officially purchased by the government at $1.29 cents per ounce, rather than the then-current market price, which hovered in the mid-forty-cents-an-ounce range. But all was not as it seemed. The government retained what it defined as "seigniorage," or half the silver. Even then, the price to producers was still 64.6 cents, or about 19 cents above the world market price. This constituted nothing more nor less than a government subsidy to silver producers. That was followed, in May 1934, by the Silver Purchase Act, which directed the Secretary of the Treasury to buy domestic and foreign silver until one-fourth of the value of the nation's monetary stock was composed of silver, or the price reached $1.29 an ounce on the world market.

In a small way, Uncle Sam also helped, when, in 1933, the government "temporarily" suspended the required yearly assessment work on unpatented claims. This freeze was in effect until 1939, when the rule was reinstated. Washington had at last become a friend to gold and silver mining. Old-timers could not believe their good fortune, and prospectors started scurrying about the valleys and hills seeking their strike. However, as the industry eventually discovered, once Uncle Sam arrived, new rules and regulations inevitably followed.

The hard times, plus the boost from federal government policies, revitalized interest in gold and silver mining. A spectacular surge in production followed. In 1933, Colorado's gold production topped $6 million, while silver languished in the $700,000 range. In 1934, gold production jumped to $11 million and then annually topped $12 million for the rest of the decade. Silver production did not reach those heights, but vaulted to $2 million in 1934 and found itself in the $3-million to $5-million range for the remainder of the 1930s.

Another effect of the renewed interest appeared in 1934. That year the number of active lode mines in the state more than doubled, to 672. It must be kept in mind, though, that not all of these were gold and silver mines.

In 1939, such old familiar mining counties as Clear Creek, Teller, San Miguel, Park, and San Juan all topped $1 million in gold and silver production, with Teller in the $5-million range. Even such familiar names as Boulder, Lake, and Gilpin Counties found gold mining at least somewhat ahead of what it had been in previous decades. The major producer, however, was Eagle County, with its Eagle Mine, a combined copper-iron-gold-silver operation, which led the state in silver production. The total Colorado precious-metal production that year surged to nearly $19 million.

Just the prospect of finding a few flecks of gold looked exciting to those who were out of work. It could be the key to getting off relief and regaining some pride and purpose. An ounce or so a month would greatly help in keeping body and soul together, so off wandered the would-be prospectors, tramping into the hills and mountains in hopes of finding enough to keep themselves and their families fed. The only problem was that many of them knew absolutely nothing about where to look, how to pan, or what equipment would be needed. Nor did they usually understand that even if the property looked abandoned, owners might be paying taxes on it and would not look too kindly on a stranger's panning for "their" gold.

In stepped the New Deal. The Public Works Administration in Denver sponsored classes in the "art of placering," with both men and women taking lessons. Once they learned the skills, many of them did not go far to prospect. Gold was being panned in Denver's streams, including Cherry and Dry Creeks, which had started all the excitement back in 1858, as well as in nearby Arapahoe and Adams Counties. Only small amounts were found, but anything helped. The mountains offered more hope, though. What the nineteenth-century miner would have thought of all this may only be imagined.

During the summer and fall, new life came to old Gilpin County districts, such as Russell Gulch, and around long-ago mining hot spots Central City, Black Hawk, and Nevadaville. Clear Creek became a mecca no matter where it wandered. Out went the men with their pans and shovels, and soon with sluices and rockers. Quickly men were busily panning and sluicing, not only there, but throughout the older gold-mining districts as well. Clear Creek and its tributaries had not seen so much work and excitement for a generation or more.

An interesting observation by a nurse at a Durango hospital during these years sheds a bit of light, describing the men (in this case, both hard-rock and coal miners): "You'd get one of the miners in. They'd been used to a lot of liquor and they were the poorest subjects for anesthetic. Took forever to get them to sleep before the operation could begin."[3] A 1937 report estimated that Colorado placer miners averaged sixty working days per season and managed to recover about $146. There were not fortunes to be made, but anything helped during those trying times. Merchants in such communities as Fairplay and Alma gained a bit of income supplying the miners, and Silver Cliff gained a moment of fleeting fame when some rich strikes were uncovered nearby.[4]

For most of the 1930s, interest in placering remained high, despite the fact that the number of placer operations decreased after the early enthusiasm wore off. In 1939, only 21 remained, employing a total of 119 miners. Larger operations continued, including dredging in Park and Summit

Counties. One dredge in the Breckenridge district, for instance, worked 808,000 cubic yards during the 1937 season. Its eighty-eight buckets probably moved more earth that season than its human placering counterparts did during several years.

Old hopes never died either. Commissioner of Mines John Joyce spurred these expectations on with enthusiastic comments. He felt "there are many well-known heavily mineralized sections of the state that have never been thoroughly prospected because of their early date, [and] remoteness which, at the present time, with modern advancement made in the means of transportation, coupled with new metallurgical discoveries and inventions, offer promising and fruitful fields."[5]

Lo and behold, the Happy Canyon placer in Douglas County (not a particularly well-known gold mining region!) briefly garnered headlines and interest, and even, according to a 1935 report, resulted in "numerous" sales of placer claims. In both old and new discoveries, however, only gold in the amount of two ounces or greater could be sold to the mint—which was conveniently located, as it had been for years, in Denver. Smaller amounts had to be sold to dental supply firms, jewelers, and others.

Despite such excitement and expectations elsewhere, the "gold king" county, Teller, consistently led in gold production. For example, in 1939, its gold mines produced more than $5 million of the state's $12.8 million total. Water problems, however, continually hampered the district, as the mine shafts went down during the twentieth century. The construction of the Roosevelt Tunnel for drainage solved the issue briefly, but by the 1930s, the mine's workings dropped below the tunnel level; thus, the costly necessity of pumping arose all over again.

In July 1939, one of the major operators in the district, the Golden Cycle Corporation, which operated mines and a mill at Colorado Springs, decided to resolve the matter with another tunnel at a lower depth. Construction started in July 1939 and was finished two years later, at a cost of more than $1 million. The Carlton Tunnel, named for one of Cripple Creek's most loyal and determined mine owners, Albert Carlton, promised relief. Bert Carlton, who had started in Cripple Creek back in the 1890s with a transportation company, got the exclusive right to sell Colorado's best coal, which came from the Colorado Fuel and Iron Company's mines. As the Cripple Creek mine shafts sank deeper and water became an increasing problem—at the turn of the century—Carlton found himself with a bonanza, serving the coal-fired steam hoists and pumps.

Thereafter, Carlton built the Roosevelt tunnel, and then, in the lean days around 1916, after the rich ore was gone, he started consolidating mines. He won some struggles with other owners and, along with his Colorado Springs mill, eventually merged them all into Golden Cycle. By 1930, Carlton had consolidated almost all of the major mines into his company and kept Cripple Creek mining going profitably. After he died in 1931, his wife, Ethel, eventually became the moving force behind the tunnel project, as the new gold price put Cripple Creek back among the country's major gold producers.

The Carlton Tunnel, six and a quarter miles long, and financed entirely by the Golden Cycle Corporation, drained the mines successfully. Completion, however, came too late to be of help, for World War II loomed right around the corner. One of Cripple Creek's old-timers summarized Carlton's contribution: "Bob Womack discovered the place. Stratton was its beacon light in the boom days. A. E. Carlton was its heartbeat."[6]

By 1941, Cripple Creek's underground shafts and works neared a length of approximately 1,000 miles, or, as one person noted, almost the distance between Denver and Chicago. It had become Colorado's major mining district, surpassing all of the nineteenth-century bonanzas, and the future looked auspicious.

Clear Creek County did not have the same kind of success, although Idaho Springs' gold mines kept the industry going, actually returning a small slice of the excitement of the old days. Silver, however, was another story, there as elsewhere. Even with a lower cost of living than in years, a reduced price for supplies and materials, and a labor surplus that pushed wages down, the expenses of deep underground mining, combined with lower-grade ore, killed whatever hopes Georgetown and Silver Plume might have entertained.

A few examples illustrate the trends of the times. The Gold Belt Tunnel, despite its name, was a silver property on McClellan Mountain. Its prospectus promised much at mid-decade, but production never matched the written blandishments. The Josephine Mine was sold for unpaid taxes in 1937, but the new owner did nothing with the property. Tax sales of mining claims and occasionally even a well-known mine occurred quite frequently, both in these years and in the next generation, when some people thought the five or so acres might make a wonderful mountain home.

Though neither the Gold Belt nor the Josephine properties had been particularly noteworthy or figured prominently in Clear Creek County mining history, the Stevens Mine had had its moments of fame—but not in the 1930s. In 1933, it was leased, and the "present work consists of cleaning out ice and snow." Then the lessees planned exploration "seeking new ore bodies," an enterprise that echoed throughout mining history. Nor did the mine's fortunes improve much after the war, going through tenants and hopes almost as fast as the changing seasons.

The *Georgetown Courier*, like its contemporaries throughout Colorado, continued to boost local mines and encourage hopes, although it no longer carried a regular mining column. Such headlines as "Silver Hopes High" did not promise production, but anything that looked encouraging merited an article. The Santiago Mine, at 12,000 feet, seemed promising in 1937. Eight miners worked there and fifty tons daily were taken to the mill for concentration. The next year's report found only the dumps being worked, although the following year, a reduced crew again was underground.

"Working the dumps"—that phrase had long been an epitaph for a mine. Granted, low-grade ore had been thrown out earlier, because it could not be profitably milled or smelted. That changed in the 1930s, when even small profits were considered worth going after. Still, salvage projects did not augur well for a mine's long-range future.

An example of a district that tried to capitalize on the mining excitement was La Plata, as prospectors/miners nosed around its canyon and mountains. The Gold King had raised hopes in the late twenties and early thirties, but drew more money from investors' pockets than from ore. The problems there, as with its neighbors, were narrow veins and only a few high-grade ore pockets.

The Red Arrow Mine, on the western slope of the La Plata Mountains, sparked a "rush" of prospectors into the district, with its "sensational discovery of an exceedingly rich" gold vein in 1933. Montezuma County, never noted as a mining district, suddenly found itself in excited headlines and in a "rush." Eager miners staked claims, blasted adits, and dug into the mountainsides. Occasional shipments of gold and silver from "seven lode mines" created brief local flurries of excitement.

Reports in the *Durango News*, on August 10 and September 7, 1934, typified the rousing stories. The "famous" Neglected Mine, on nearby Junction Creek, started shipping ore and rated publicity as "one of the most encouraging pieces of mining news to break in this section in recent years." That announcement meant "much to Durango," because "money brought" into town "in supplies and wages," the editor enthusiastically predicted, would be in the range of "$10,000–20,000 per month." As with so many projects in bygone days, the Neglected neither became famous—perhaps infamous for disappointing investors would be a better description—nor did it bring in anything close to that amount of money or jobs to Durango or La Plata County.

Two events did remind old-timers of years gone by. A 1937 news story forecast that a "modern flotation plant" would be built in the canyon; there was nothing like a projected new mill to get folks stirred up about coming prosperity. (The "modern plant" never saw the light of day, by the way.) Meanwhile, the year before, another story had reminded people of the dangers of mining, when a snow slide killed six men and a woman at the Doyle mine.[7] Like the coquette it had always been, La Plata Canyon teased and promised but yielded little. Each excitement briefly rekindled golden hopes and, if nothing else, provided a few jobs and purchases in the otherwise depressed days of the 1930s.

With support from the New Deal, unions throughout the country gained new power and influence. This led to a few tense days for Silverton, Charles Chase, and the Shenandoah-Dives mine. An extraordinary talented and skillful individual, Chase had come to Silverton in the late 1920s from Telluride to manage the mine. Just as he finished building a boarding house at the portal, a two-mile tram, and an underground crushing plant, the crash of 1929 hit. Suddenly, with prices of lead, copper, and silver collapsing, only gold kept the mine open.

Realizing he could not keep going, and knowing how much the mine meant to Silverton, he asked his 150 miners to agree to a 25-percent wage reduction. They did. Then he called on local merchants and landlords and asked them to voluntarily lower prices. In both cases, he explained that without help the mine would close, and Silverton's plight would be worse. They agreed, as did some out-of-town creditors after a bit more persuasion. Chase laid no one off and fed out-of-work "tramp" miners, who hiked to the high mine, a meal even when he could not hire them. In 1935, he restored the old wage scale. His mine, the Colorado Bureau of Mines reported in 1939, was "up-to-date in every detail." On top of this, Chase was one of the few pro-union mine managers in the region.

Despite all these things, a bitter episode broke out in 1939, when the CIO called a strike for higher wages. Meetings and explanations solved nothing. Finally, fed-up local miners ran the organizers out. After a tense and confrontational episode, Chase won the day and kept the mine and mill operating into 1953.[8]

December 7, 1941, ended an era in Colorado mining. On that "date which will live in infamy," the attack on Pearl Harbor changed mining everywhere. Colorado miners were already leaving to take higher-paying jobs elsewhere, particularly in the West Coast defense industry. As the country and state confronted the military crisis, mining would be called upon to play a strategic role—but with its base-metal and coal mines, not precious-metal gold and silver mines.

As war descended on Colorado in 1941–1942, The *Colorado Year Book* took time to pause and consider some of mining's contributions to the state, besides the wealth that had poured forth since 1859. Thanks to the industry, the mountains, long serious barriers to transportation, had been penetrated by road and rail, and water had been harnessed for hydroelectric power. Cities had been born and matured; some had died. Industry and agriculture had taken root and profited for decades, and several generations of Coloradans had found work in the mines and mining-related industries. Even tourism had been profitable. It had been an exciting run, even in the dark days after World War I. Now the industry came to a watershed as another world war broke out.

The Colorado mining recession, which dated back to the late teens, eased as the United States went to war, but the revival was not for the precious-metal industry. Regulation had been a fact of mining life since the early New Deal, and now gold and silver miners found out what government oversight could mean.

A series of government fiats overwhelmed the industry. Even before the United States entered the war, the Metals Reserve Company, created in June 1940 to build stockpiles of "strategic and critical metals and minerals," paid premiums for copper, lead, and zinc. This helped some fortunate gold miners, who mined those substances as byproducts, to make a profit. Then came Order P-56 (March 2, 1942), which stated that any mine, whose value in gold, silver, or both exceeded 30 percent of its total output, would "no long[er] be entitled to a priority rating." That deprived virtually all Colorado gold mines of the priority rating needed to purchase essential equipment and supplies. Many operators closed their mines before the next shoe dropped, because the industry already faced higher taxes, a loss of skilled miners, and equipment shortages.

Then came the final blow—infamous, in some miners' view. Gold Limitation Order, L-280, which went into effect on October 8, 1942, closed nonessential mineral mines by not allowing operators access to replacement parts or materials or to "break any new ore, do any development work, or start any new operations." Furthermore, no ore or waste, "either above or below ground," could be removed. Nothing could be done "except the minimum amount necessary to maintain buildings, machinery, and maintain equipment in repair." Gold mining had been defined as "nonessential" to the war effort, with one exception. Mines that had produced less than 1,200 tons of commercial ore in 1941 could continue in operation, provided they did not exceed 100 tons per month.

As if that were not enough to stop the precious-metals industry in its tracks, skilled miners were shunted to the essential mining segment of the industry. The government gave them occupational deferments as long as they stayed in nonferrous mining jobs; if they moved elsewhere, the miners lost those deferments. The War Manpower Commission additionally ruled that no former gold miner could "obtain a new job west of the Mississippi without the approval of the United States Employment Service."[9]

Though miners and others argued, for a while, that gold was needed to pay for the war, patriotism carried the day. Other mining was more significant to the war's outcome. Dredging and most small operations closed immediately. Operators in Colorado's top district, Cripple Creek, appealed the date and were given a short reprieve, until June 8, 1943, to finish current projects. Then gold mining stopped, except for maintenance. Silver had never been a factor in all the discussions, as it had become primarily a byproduct rather than a principal ore. While other Colorado mining flourished during the war, both in production and exploration, inspectors' reports simply listed production as "none" in gold and silver properties. Although some gold and silver were produced as byproducts, Colorado precious-metals mining had gone to war.

In July 1945, with the war nearly over, the gold and silver mining industry was finally released from government sanctions. No boom followed, however. For a while, into the early 1950s, placer production briefly prospered, thanks to a large dredge operating near Fairplay, but in January 1952, the dredge ended its run. Low gold value and increasing labor, supply, and equipment costs all combined to do it in. That would become a familiar pattern in postwar mining.

Underground mining was hurt initially by a shortage of skilled miners and equipment, a blow reflected in the fact that gold and silver production sank to lows unequaled since the 1870s. Indeed, much of what was mined came as a byproduct of base-metal mining, particularly out of the Red Cliff District in Eagle County.

In the years that followed, the same problems continued to plague the precious-metals industry. Interest in mining only gold or gold/silver understandably waned. With the price of gold regulated, owners could not compensate for the increasing expenses of labor and materials, unlike with base metals. The days of high-grade deposits—the one factor that might have stirred interest—had vanished almost entirely years ago.

A federal study dramatically showed what had happened. Gold held the top ranking from 1858–1873, when silver replaced it. Then, thanks to Cripple Creek, gold regained the number-one ranking from 1897–1942. After that, zinc grabbed the honor and held it into the 1950s. In 1941, gold and silver represented 77 percent and base metals 23 percent of the total value of Colorado's five long-time significant metals: gold, silver, lead, copper, and zinc. By 1951, there had been a startling reversal, with gold and silver constituting only 17 percent of the total.[10] Even that would soon change, as molybdenum and uranium surged to the front of the pack.

How bad had the situation become? In 1958, Lake County, for the first time since the Oro City rush in 1861, reported no mines operating after a strike closed the Climax molybdenum operations. Low prices across the board were forcing the closure of base-metal mines throughout the state, and with them went their precious-metal byproducts.

Despite all this gloom and despair, one long-lasting concern had been answered: a modern, up-to-date mill had been built. The Golden Cycle Corporation constructed one in the Cripple Creek district in 1950–1951, even as local mines shut down or worked only on development. When the mill opened, gold mining revived, thanks primarily to Cripple Creek, and topped $4 million. Teller and San Miguel led the way in producing more than 80 percent of that total, with the latter yielding much of it, once again, mostly as a byproduct from the Idarado mine, which led the state in base-metal production. Even with the Cripple Creek resurgence, 45 percent of precious-metal production came as a byproduct.

While mining struggled, the state celebrated the centennial of the Pike's Peak rush in 1959 with pageants, ceremonies, and tributes to the "stalwart" pioneers. The "Rush to the Rockies" centennial brought in tourists and profits, but no real understanding of the state's mining heritage or of the people involved. Gold mining had given birth to the state, but few Coloradans now worked in, or made a living from, the precious-metals industry. The words on the drop curtain of Denver's Tabor Grand Opera House seemed prophetic: "how fleet the works of men."

A far better tribute had debuted three years earlier at the Central City Opera House, near where it had all begun a century before. Douglas Moore's opera, *The Ballad of Baby Doe*, caught the spirit of Colorado, mining, and the era, in a manner the written word could not. The Tabors once more became front-page news.

In a curious coincidence, the Tabors rose to fortune and fame on silver, which was gaining new life. The hopes of the silverites of 1893, and of

William Jennings Bryan in 1896, seemed about to be realized. In 1961, the federal government freed silver from government regulation and the price soon rose to $1.75 per ounce, the highest in Colorado's history. For years, old-timers had been fond of predicting that $1.30 an ounce silver would bring back the great days. Did the rebirth really occur? No, no new Horace Tabors found their Matchless mines, and hopes went a-glimmering out once more. That price was not enough to offset the expenses, which had risen steeply since Horace's day. No miners and prospectors roamed over and dug deep into the old silver districts. Only sightseers continued wandering about and wondering what had transpired there.

Colorado poet laureate Thomas Hornsby Ferril perfectly caught the ambiance and the desolation in his poem, "Ghost Town":

> Dig in the earth for gold while you are young!
> Here's where they cut the conifers and ribbed
> The mines with conifers that sang no more,
> And here they dug the gold and went away,
> Here are the empty houses, hollow mountains,
> Even the rats, the beetles and the cattle
> That used these houses after they were gone
> Are gone; the gold is gone,
> There's nothing here,
> Only the deep mines crying to be filled.[11]

Just when things seemed to be looking up, Colorado received another jolt: Both the Golden Cycle mining and milling operations stopped operations in 1961. Cleanup operations the next year ended an era at Cripple Creek. The familiar reasons were proffered: increased costs across the board made it "almost impossible to mine and produce gold." Thus closed "Colorado's last gold camp."[12]

It had been an amazing run. Through 1961, Cripple Creek mines had poured out more than $424 million, ranking it as Colorado's greatest gold producer and the second largest gold district in the United States. All that remained were the Molly Kathleen and El Paso mines, which remained open to mine the tourist trade. Colorado mining faced a stiff challenge in the 1960s, which Governor Steve McNichols clearly described in a 1962 address to the Western Resources Conference meeting at Golden:

> Here in Colorado the challenge to develop our mineral resources is an ever-continuing one. Colorado began mineral resource development after the discovery of gold by John Gregory near here in 1859 and the develop-

ment of minerals has been synonymous with the growth of Colorado for more than 100 years.

The governor might also have mentioned a challenge that was only peeking over the mountains in 1962: environmental issues. This subject was not completely new, of course. After all, Rossiter Raymond had raised some of those concerns back in the 1870s.

The problem had been around for years, as shown by a 1909 lawsuit that involved a homesteader near Creede and a company that discharged waste into a stream. The mining interests lost that case, though they claimed that the industry was absolutely dependent on discharging waste material into the stream. Interestingly, they also advanced the idea that in time of water scarcity, the state constitution gave mining priority over agriculture.

The matter came to the forefront during the depressed 1930s in a major fight between mining and agriculture. At issue was the water quality of the Clear Creek drainage, which encompassed the once-famous Central City and Georgetown Districts. After passing out of the mountains, the creek drained into the South Platte River, where it was used for drinking and public use. It eventually reached northeastern Colorado farmland, where the "mineral enriched" water (some members of the mining industry thought that was a bonus!) was used for irrigation.

During the 1930s, though, small mining operations proliferated in old districts, which made the conditions worse. Operators with little money and less environmental concern made no provisions, or only weak ones, for impounding their tailings, which ran directly into the stream. A report published after an examination of mines along the stream summarized the situation with these terse comments: "no provision for impounding, sands and slimes flooded North Clear Creek, dump into stream, ponds not effective."

As noted earlier, Colorado had had statutes outlawing stream pollution since territorial days. Now mining companies found themselves in court, the key case being *Wilmore v. Chain O'Mines, Inc.* The Chain O'Mines was a late 1920s project that had combined numerous overlapping claims above Central City into one large-scale operation. Mining up from the older, lower operations, through controlled blasting, the company eventually caved in the mountain and created what became known as the "Glory Hole." Milling everything in sight, the company discarded astonishing tonnages of waste, which became the central figure in this cause célèbre. The plaintiffs, who owned farms in downstream Jefferson County, claimed that the defendants had discharged "mill tailings and slime" into the stream and onto their land, thereby harming their property.

The defendants responded with a combination of confidence and desperation. The companies' defense, firmly rooted in nineteenth-century attitudes, showed heavy-handedness, candor, shrewdness, and ingenuity. After admitting to depositing tailings and slime into Clear Creek, they stressed these main points.

1. The tailings were not injurious.
2. They had been discharging them at the time the farmers acquired their lands (a "coming to the nuisance" defense).
3. Usage and custom of many years had permitted deposition of such materials into the watershed.
4. The cost of developing a suitable deposit place outside of the streams was in excess of profits of the operation.

Why had the plaintiffs not complained earlier? Had they not thus acquiesced by their silence?

The plaintiffs pointed out that above the Chain O'Mines' mill, and at other plants, the stream was clear "most of the year," but below the tailings it was a "thick muddy flow that at times approached the consistency of thin cream." This damaged irrigation, harmed livestock, and ruined crops. The Jefferson County District Court issued an injunction that "fully and permanently enjoined and restrained [Chain O'Mines] from discharging slimes or tailings into Clear Creek or its tributaries." The Colorado Supreme Court upheld that decision, "notwithstanding 67 of the most prominent attorneys in the state" asking the court to reverse the lower court. The decree was issued on April 12, 1935.[13]

Chief Justice Charles C. Butler dissented from the majority opinion and well summarized mining's position: "The holding in this case, if adhered to[,] would seriously and unjustly cripple the mining and milling industry." To mining, he continued, Colorado owed its birth, its financial salvation during the 1893 panic, and a "large measure of its prosperity at all times."

The *Engineering and Mining Journal* (April, 1935) tried to find a positive result, pointing out that "the thing enjoined is pollution," not the mining or the industry. It also pointed out that "conference and compromise" offered a better solution, "if the contestants are not too obstinate." That was a big "if" with this industry.

In its November issue, the *EMJ* came back to the question in an editorial entitled "Tailings Disposal and Stream Pollution." The editor, after visiting the district, noted the obvious extra cost burden that "has become a major concern to some operators." He observed, though, that "few . . . are inclined to question the essential justice" of such regulation. One had only to look at the "plight of the farmer who finds slime-burdened water ill suited for watering stock or irrigating land."

The editorial concluded with a warning that Colorado mining should have heeded but did not. Conditions in Colorado "offer no exception to the growing sentiment against stream pollution" across the nation. The journal expected Congress to enact a nationwide measure to prevent stream pollution in the next session or soon thereafter. Meanwhile, Crested Butte's *Elk Mountain Pilot* added that the decision was important to fishermen, as streams would become "clear enough for fishing."

With the general public becoming more aware of environmental impact issues, much confrontation and litigation appeared to be in the offing—but they did not have to be, as Charles Chase showed in Silverton. As the superintendent of Silverton's Shenandoah-Dives Mine and Mayflower Mill, he planned for pollution control. Whereas every other mill in the San Juans discharged tailings into the "nearest available stream," Chase's did not. Explaining why, he wrote: "Because of my personal repugnance for the lack of consideration of the public interest involved in this practice, I undertook to withhold from the Animas [River] the tailings of our new project."

Depression-wrought woes eventually stopped some of the projects he had in mind, but Chase did respond to downstream farmers' complaints in 1935 by installing tailings ponds. Supported by stockholders, he did the best he could. "I have always remembered an expression of one of our largest owners in our early days: 'It is a shame to spoil such a beautiful stream.' "[14]

These two variant responses exemplified the directions mining was going. Confrontation and litigation on the one hand, and an unfortunately rare combination of caring stockholders and a concerned superintendent on the other, promised interesting years ahead for an industry already in decline.

A 1964 report on the mineral resources of Colorado painted a realistic picture of Colorado mining, which had been in a "general decline" since the end of World War II. A few familiar mines still operated. The Camp Bird, Sunnyside, and Idarado Mines were active in the San Juans, thanks to base metals; mining in Rico concentrated almost solely on pyrites to make sulfuric acid for uranium milling. Yet, with the "silver lining" hope that kept mining going, the report concluded, "Despite the current trend, the future of metal mining in Colorado should be thought of in terms of moderate optimism."[15]

Some of the optimism was founded in Creede, where the improved price of silver somewhat rejuvenated the district. The Homestake Mining

Company came down from the Black Hills of South Dakota to open the Bulldog Mine, and its output, joined by that of a couple of others, put the county into the range of several millions of dollars per year during the decade. This is what the industry had hoped for, but it simply had not materialized as they expected. As 1969 ended the decade, Colorado gold production barely topped $1 million and silver $4 million. This hardly equaled the production of a single prospering county in the glory days. Three old-timer counties (Mineral, San Juan, and San Miguel) accounted for more than 90 percent of that amount, while counties like Gilpin, Lake, Pitkin, and Teller mined none, and Clear Creek yielded less than $7,000.

Ahead lay a new mining world, behind were the remains of an era now gone. Back in the fifth century B.C., the Greek philosopher, Heraclitus, observed that "Nothing endures but change." The late twentieth-century Colorado mining industry was going to find out how true that was.

19

Mining on the Docket of Public Opinion: The Environmental Age

Almost from the days of the Pike's Peak gold rush, there had been Coloradans and others concerned about mining's impact on the environment. The reasons have been many, but the discussions and actions generally were local. Rossiter Raymond raised the issue of the destruction of trees in a national forum, but Rico dealt with the problem at its immediate community level.

Sitting beside her adored O-Be-Joyful Creek, in the heart of the Gunnison country, Helen Hunt Jackson mused about the effects of mining. To her, the sparkling stream and the nearby field of purple asters were far more valuable than the minerals for which the prospectors searched and the miners dug. "There is no accounting for differences in values; no adjusting them either, unluckily."[1] Central City in the 1860s, and Telluride in the 1890s, became concerned about mining pollution in their water sources, but, faced with the prospect of harming their economic base, chose a prosperous economy over public health.

Lawyer, historian, and environmentalist James Grafton Rogers, in the mid-twentieth century, wrote about his beloved Georgetown area in *My Rocky Mountain Valley.* One of the themes was how mining had affected the land:

> Of all the scars men leave on nature the mine dumps in the West are the most conspicuous and permanent. There is a diagonal band across Colorado, from Boulder to Rico, perhaps fifty miles wide and two hundred and fifty long, where men found precious metals. . . . [T]hese relics of the miner seem unchanged although a century has gone past. The scars do not heal.
>
> The mine dumps, the piles of rocks excavated from shafts and tunnels, the beds of mill tailings all up the valley, the heaps of boulders left by gold dredges about a generation ago along the canyon stream—these seem untouched by time.[2]

Anyone near a smelter could see the smoke's impact on the land, but not so clear was the impact on people's health in Leadville, Central City, or Colorado Springs. Nor was this concern limited solely to Colorado. J. Ross Brown, John Muir, and Mary Hallock Foote, among others, commented on the environmental problems created wherever mining was or had been.

Several concerns had been expressed—and litigated—in Colorado earlier in the twentieth century, but no national or even statewide anxiety had yet been exhibited. All that changed in the 1960s, when the storm that had been gathering for years finally broke. The great environmental awakening was at hand and a popular cause was a-borning. No Colorado incident brought this about; a multitude of factors figured into the equation.

The publication of books such as Rachel L. Carson's *Silent Spring* and Harry M. Caudill's *Night Comes to the Cumberlands* popularized the subject of the environment and frightened readers. Increased knowledge about the environment and about industrial impact on it intensified the public's consciousness of the threat that pollution posed to all Americans. Also, during the administrations of Presidents John Kennedy and Lyndon Johnson, the federal government became more actively involved with environmental matters. This, combined with growing activism among America's youth (who were particularly cause-oriented during the 1960s and 1970s), meant that Americans soon became polarized over environmental issues.

The result was, among other things, a host of federal laws passed during those decades dealing with environmental issues. Mining crossed a watershed in its history then, though the long-range significance of this shift in public perception and concern went unacknowledged at the time. It had been a long time coming, but the effect would not be totally recognized until later.

The nineteenth-century industrial attitudes and philosophy about progress, profits, and prosperity had carried through the first six decades of the twentieth century. Now, in the latter half of the twentieth century, environmentalism was "in" and the day of judgment had come. *Conservation*, *ecology*, and *environment* were words that had carried little meaning a generation earlier. Now, they suddenly became the watchwords of a new and increasingly popular mindset.

The mining industry, nearly blindsided, found itself on the defensive, as terms such as *raper*, *polluter*, and *exploiter* were lobbed at it with reckless abandon. Some members of the industry responded in kind, dismissing their critics as "stupid idiots," "socialists," "commies," and "loudmouth hippies." Caught off guard, the industry responded in a variety of ways, which ran the gamut from digging in its heels on the issues to being willing to talk to and work with environmentalists.

It was probably no accident that all this occurred at a time when Colorado mining (including gold and silver mining) had already sunk into the hardest times in its history. Far from being the economic mainstay of the state, mining was in a precipitous decline, apparently contributing little besides damage to spectacularly beautiful portions of the state. "Spaceship Earthers" peered about Colorado and saw plenty of problems. Tailings ran into streams and rivers throughout the mining regions: Clear Creek, the Animas River, the Blue and Swan Rivers, the Gunnison River—the list stretched on and on. Where they dug, miners left behind water draining out of mine portals, some of it highly acidic. The Red Mountain District in the northern San Juan Mountains was a classic example of this problem.

Dredges had left behind their "dung," mines their dumps, and smelters their tailing piles on mountainsides and river valleys. Some people thought them historic parts of the state's heritage. Far more considered them eyesores and environmental disasters. Mining roads, cut into the tundra around mountains and through valleys, had caused erosion and environmental damage that would take decades to recover from. Tree stumps and denuded hillsides further marked the path of mining, along with railroad grades, cuts, and collapsed bridges.

Miners, like their contemporaries, had also littered; wherever they had worked and lived, they left a trail of rusting equipment and debris. Corroding tin cans, broken glass, fallen-down buildings, and a myriad of other discards and garbage became another heritage, even after World War II scrap drives cleaned up the accessible sites. These remains marked the passage of an industry, its people, and an era.

To many Coloradans, the very word *mining* conjured images of the "rape and run" strip-mining and open-pit operations that had created such horrors in the coalfields and copper mines of both East and West. Seepage discolored creeks and rivers miles from its source at a mine's portal. In fact, coal-mine drainage threatened nearly the entire upper Ohio River drainage. These examples raised an obvious question: Was Colorado gold and silver mining any different?

It certainly was, but time and patience would be needed to inform and educate the general public. Denver's *Mining Record*, one of Colorado's leading defenders of the industry, took up the cause through its cracker-barrel philosopher, Prunes the burro. Environmentalists, politicians, and a host of anti-mining folk got "kicked" repeatedly, and hard, as Prunes brayed loudly:

> Easterners will believe anything unbelievable about the West.
>
> If the old-time miners had been forced to replant the trees and restock trout streams like today, they'd never have made it!
>
> An honest politician is one who, when he is bought, will stay bought.
>
> I want a front row seat when the Environmentalists start digging for coal to satisfy their energy needs.
>
> I'm lucky, my food is grass and bark and leaves. No bureaucrat can regulate their growth or prices.
>
> Ask not what the government can do for you—but of what the government can do to you.

Equally opinionated were some of the bumper stickers (Americans had a penchant for them in the 1970s) that proclaimed the miners' stance in a few pithy words: "Ban Mining. Let the Bastards Freeze to Death in the Dark." A gentler, but equally pointed and uncompromising, sticker reminded readers: "If it can't be GROWN, it has to be MINED." There was one that simply said, "I Support Colorado Mining" (or substitute any other mining state). Another positive one declared, "Mining Is Everyone's Future." Thus both sides made their statements as they passed each other on the streets and highways, though it is highly unlikely that anyone at all was convinced to change an opinion.

This automotive show of support might have made Prunes and his supporters feel better, but the tide of popular opinion ran against them. More federal laws and regulations, along with Colorado ones, appeared as the century raced toward its conclusion. Federal agencies, such as the Environmental

Protection Agency (EPA), enforced these laws, sometimes with too much vigor, according to mining Coloradans. Retrenching somewhat from its hard-line position, the *Mining Record* (February 14, 1979) expressed hope that through education, conservation, enlightenment, and the sound management of resources, a balance between environmental preservation and public needs could be struck.

Mining conferences in Colorado and throughout the country grappled with environmental questions. The *Mining Record*, at a special National Mining Conference, on February 9, 1973, proclaimed, for example, that "Thou Shalt Not Pollute" had become the eleventh commandment. A report from the Western Resources Conference, held in Golden two years earlier, was a little more realistic but still emotional: "What is needed is science and technology to match the fervor of ecology ranters and demonstrators, with solutions, remedies, and changes that will work effectively." The report authors suggested that concern be "centered on practical and realistic" methods of "solving environmental problems." Division by "faddism and fomentation" only made matters worse.[3]

Into the fray rode defenders of mining, especially during the 1970s and 1980s. Grand Junction's *Daily Sentinel* pointed out that tourists "love the remains of mining days," which could not be classified as "pollution." "Cameras are pointed at the 'pollution,' in most cases with the 'scenery' used as background." The *Mining Record* complained that "unrealistic laws and regulation" made it "difficult" for all mining companies, whether small or large. "The mining industry has always been faced with many diverse challenges; recently additional challenges have been placed on the industry . . . government land and environmental regulation."[4]

The *Denver Post* (January 4, 1981) tried to balance the picture of "Huns and miners being practically synonymous in some quarters." It pointed out that industry members had brought this "on themselves" through disregard of the environment. As the 1980s opened, though, mining companies began making serious efforts to reverse their poor public image through more environmental concern and action.

Talking, orating, letter writing, arguing, complaining, and publishing articles in newspapers and popular magazines kept the issues on the front page, or certainly in the public view, but did not solve or even touch upon some of the basic issues. There were tens of thousands of abandoned mines and prospect holes scattered around the Colorado mountains. Who was going to pay to reclaim them? Most of them had been abandoned long ago, and the former owners and stockholders were dead, which left nobody to sue or carry out reclamation. As for currently owned but non-operating areas, was the present owner liable for the actions of previous owners? Were current stockholders liable for damages caused by their companies years earlier? Was the Colorado public liable because the mines were within the state boundaries, or was the federal government at fault—or at least complicit—for issuing the owners their patents?

What about open portals that attracted unwary youngsters and meandering hikers into the mines, not knowing where a shaft might be, or a winze or stope to stumble into, or rocks ready to tumble down on intruders, or timbers ready to break? Some mines contained dead-air pockets, and a few harbored gas that would quickly overcome anyone who wandered into these death traps. It was a legal nightmare and a costly future for whoever was eventually held responsible.

Keep in mind that gold and silver mines were only a small part of the national problem—not that this fact made people feel a great deal better. Coal mines, copper mines, and others also polluted, often in a blatant fashion that aroused the public; uranium mines were particularly egregious offenders. Coloradans faced some of these problems with its uranium and coal mines, especially along the eastern foothills, where some subdivisions in Boulder and Jefferson Counties, for instance, squatted on top of unmapped coal mines. Western Colorado was littered with the remains of the 1940s–1960s uranium excitement, which left behind mines, dumps, smelters, and dying miners, along with some scattered radioactive "hot spots."

In some Colorado locations, the industry faced no-win scenarios. In 1976, Allan Bird, superintendent of Silverton's Sunnyside Mine, put it in perspective, when his company encountered opposition to the destruction of trees for expansion of a tailings pond: "The EPA people looked at it. The water quality people looked at it. They all approved the plans. It was a situation where it was the choice of preserving 30 acres of willows or putting a town out of work." Vocal critics thought that was no excuse for destroying the willows—or the land that the mine company owned.

Silverton's old Sunnyside Mine had gained new life when operations resumed in 1961, and had become Colorado's largest gold producer by the early 1970s. Standard Metals placed its mine in full operation, working six days a week to produce silver, gold, and base metals. In 1973, for example, it yielded nearly $5 million. Bird and his staff faced environmental issues, however, that would probably have driven old-time managers over the cliff. In June 1974, the snow runoff washed about 100,000 tons of "grey slime" off the dump and into the nearby Animas River. In stepped regulators and agencies; only after hearings and studies were concluded, a shut-down for repairs completed, and a fine levied and paid was mining resumed.

Four years later, the company suffered a major accident with serious environmental impacts. Working a profitable gold stope, mining moved upward toward the bottom of Lake Emma. Gradually, the miners noted seeping water trickling down the shaft. The seepage became a flood on June 4, 1977, when the lake broke through into the workings, shooting thousands of gallons of water and millions of tons of mud out into the valley and then into the Animas River. Along with that mess, it carried along timbers, mining equipment, and anything else that could be flushed out. The only extremely fortunate thing about this disaster, "one of the most dramatic mining catastrophes" in San Juan mining history, was that it happened on a Sunday afternoon when everyone was home.

The shocked residents of Durango and Farmington, forty and eighty miles downstream, saw the mess race by. It would take two years of demanding, difficult, dangerous work to clean it up and get the mine reopened. Neither Standard Metals nor the Sunnyside Mine would ever be the same again. They sold the mine to Echo Bay in 1985, but that company could not make work there profitable either. The Sunnyside—the last major operating mine in the San Juans—eventually closed in 1991.

For a while before it finally closed in the late 1970s, the Idarado mine, whose workings tunneled from Red Mountain through the mountains to Pandora, near Telluride, where the mill was located, gave the San Juans two major operating mines. Although it yielded both gold and silver, its main products were zinc and lead. The Colorado Division of Mines' 1979 report observed that the Idarado "simply continued to retain a maintenance crew," with some development work being done. San Miguel listed no gold or silver production that year, compared with, for example, more than $2 million in the last year of full production in 1976.

What got Telluride people all upset, once skiing replaced mining as San Miguel County's principal industry, was the dust blowing off the tailings pile, just outside of town up the valley toward the mill site at Pandora. This led to a joint environmental cleanup effort by the company and the town that covered the long-stretching dump with vegetation.

Over in Mineral County, the Homestake Mining Company operated its Bulldog Mine and mill at full capacity. By 1979, the mine was Colorado's largest silver producer, showing more than $14 million worth of silver and employing a crew of nearly 170 men. To the shock of Creede, Homestake closed the Bulldog in 1985. The company blamed its failure on not finding new ore bodies, the high costs of mining, and the fluctuating price of silver.

Meanwhile, the federal government stepped out of the gold market and allowed gold to seek its own price level on the world market. The removal of the government cap on gold (and the price fluctuations on the open market) sent a new jolt of potential throughout Colorado. As of December 1974, for the first time since 1934, Americans could legally buy, sell, and own gold.

Understandably, this renewed some interest in gold, but there were other factors that had to be considered before venturing into late-twentieth-century precious-metal mining. Miners' wages had risen steadily, with some earning $100 or more per day. Also, fewer skilled miners remained in the state, a problem that had bedeviled the Homestake operation at Creede. Costs of supplies, equipment, and transportation had continued to climb as well, and Uncle Sam's regulations were omnipresent and expensive to satisfy. The high-grade gold and silver ore of the early rushes had disappeared, and only the lower-grade rock deep in the earth remained in Colorado's mines. Taken together, all these factors placed the miner in the age-old bind of higher expenses versus declining ore value. Furthermore, once the Carlton Mill closed in 1962, ore had to be shipped to Canada, Mexico, or El Paso, Texas, to be refined, adding further costs to the already chancy proposition called mining. All this weighed heavily against any new rush. Also, in this situation, the small miner had little chance. The big corporations, with skilled geologists, more equipment (including the latest technology), and the financial resources to withstand initial losses to gain eventual profits, continued their domination.

Colorado's centennial year, 1976, illustrated the new era of gold and silver mining. Only the increase in price of the former accounted for an increase in value mined ($14 million-plus), not a corresponding increase in ounces of silver. Gold production had actually decreased from previous years to $5.4 million. Production would not even have been that high except that gold and silver generally came as byproducts from mines that produced other minerals. Compared to crude oil's $376 million, coal's $143 million, molybdenum's $183 million, and even sand/gravel's $41 million, gold and silver were bit players on the total Colorado mining scene. The precious minerals industry had fallen far.[5]

The largest gold producer in the last years of the twentieth and the first years of the twenty-first centuries was the Cripple Creek & Victor Gold Mining Company. In the mountainous area between the two towns from which it took its name, the company conducted extensive exploration, including geological studies, drilling, and construction of a model of the deposit to determine whether it could be mined efficiently, safely, and profitably. Environmental issues, no longer in the background, had to be studied and a reclamation plan prepared. The drilling disclosed that the district still contained large quantities of low-grade gold-bearing ore, previously seen as uneconomical for mining and milling. The company decided that an open-pit operation would be the most feasible, but that quickly raised red flags. Open-pit mines were not popular anywhere in Colorado.

Because of those concerns, Cripple Creek & Victor had to work its way carefully through numerous federal and state agencies. Such issues as miners' health and safety, the impact on the communities (Cripple Creek and Victor), air and water quality, archeological surveys, and blasting and processing procedures had to be studied and reports prepared. Finally, the company had to pay for an outside third-party review, secure its mining permits, and post a bond ($53 million on the Cresson Mine alone), payable to Colorado to ensure that reclamation would be accomplished not only on the present working site but also on older mines owned by the company. Environmental protection proposals were submitted and public meetings were held to inform—and discuss and sometimes argue. This obviously was not the way that Stratton and his friends had approached mining back in the 1890s in the district.

Nor would Stratton, or his contemporaries, have recognized all the new ins and outs of a mining operation. Following a "thorough geologic investigation," feasibility studies, an "extensive permitting process with local, state, and federal agencies," and posting of a reclamation bond, actual mining could finally began. In this modern style of mining, location drilling was conducted initially to confirm gold content. Then, after blasting, the muck was surveyed and marked with flags indicating gold ore. Non-gold-bearing rock would then be backfilled in previously mined areas or moved to a storage site.

At this point, old-timers would have been able to recognize the procedures, if not the magnitude of the operation. Large excavators loaded rock into 300-ton-capacity haulage trucks, which delivered the ore to the crusher to be ground into smaller pieces in preparation for the next stage. After placing the crushed rock on a leach pad, "like a large bathtub," a "very weak, alkaline based, sodium cyanide solution is dripped on the rock." This slowly dissolved the gold, which percolated to the bottom of the pad. Thus captured, the gold-bearing ("pregnant") solution then drained into the recovery building to await the next process. Using a carbon-absorption process, a gold-rich "mud" was heated to separate the gold and silver from any nonmetallic substances. Finally, the gold-silver mixture went to a specialized refinery for final separation.

In 2009, the company, with 300 employees, was the only major mine operating in Colorado, as it has been throughout the first decade of the twenty-first century. Cripple Creek & Victor's operation included more than just mining a mountain away. The company was sensitive to various problems, including historical and archaeological research, roads and transportation, and the aforementioned environmental issues. Trails for hiking and riding were built, historical exhibits and a history trail developed, and buildings refurbished in Victor. Archaeological work was conducted on threatened sites, a state highway was moved, and taxes were paid to Teller County. All of these things helped underwrite a variety of county and community projects. These efforts showed a great appreciation for mining's history, the community and county, and the environment. Still, there was another side to the story. Several old mining campsites disappeared, including Altman, made famous in the Cripple Creek strikes. Also gone was the mountain it sat upon.

Even if Stratton had taken all this in stride, he probably would have been stunned by another development within the industry. It took a long while, but women finally went underground in mining, as well as working above ground in the industry. Our friend Prunes made one of his pithy comments about it, "I wonder where the Tommy knawkers were hiding when WIM went underground?" Some of these women "pioneers" had organized themselves into Women In Mining and were active there, as well in as historic preservation and helping the public to understand and appreciate Colorado's mining heritage.

It was not easy being a pioneer. One young lady, who was the first to work in Silverton's Sunnyside Mine, said that she remembered facing passive—and sometimes active—hostility from her male coworkers. A couple of male coworkers quit because of that long-held superstition of its being bad luck to have a woman underground. Others made remarks about women not being able to "carry their weight." Sexual and other scribbled comments appeared here and there within the mine, and gatherings in the "dog house" for a meal could be quite tense. Finally, her fellow workers grudgingly accepted her. It took a while, but women mining engineers, geologists, and others eventually found work in Colorado mines, both above and below ground.

By the twenty-first century, women were found in a variety of mining-related posts. For example, a report on the Cripple Creek & Victor Company pointed out that "[w]omen comprise an integral part of the mine work force, not only in the office but in the mine as equipment operators and in many technical, professional and engineering positions." Progress over the past generation had been steady, albeit in an industry that continued declining within the state.

Denver, however, was revived as a national and world mining center in the last decades of the twentieth century, as companies used it as headquarters for mining operations throughout the world. While this brought mining people to the city, it did not benefit state mining except in a few instances. The Colorado Mining Association continued to call Denver home, as it had for more than a century. Its annual meeting provided the opportunity for the public and mining folk to become more informed about a wide variety of developments within the industry. The Association was also the mining spokesperson at the state legislative sessions and promoted mining in various ways throughout the year, including teachers' summer institutes and publication of the *Mining Record*.

Despite the predominance of corporations in mining and in mineral exploration, Colorado still offered a place for the "small" miner. Tom Hendricks, mining at Caribou, proved this with his successful Cross Mine, just east of the old town site. Mining in environmentally active and sensitive Boulder County took patience, concern, and determination. As Hendricks pointed out, "Mining is not a get-rich-quick scheme anymore. Those days are over. But with hard, dedicated work, you can operate one mine as an ongoing business." Hendricks conducted tours for schoolchildren and others through his operation to introduce them to modern mining and its concern for the environment. "Even though you have environmental restrictions, you can still work within them and produce," he stated.

Hendricks was the prototype of the environmentally concerned modern miner. Unfortunately, and tragically, Colorado also presented a terrible example of the old-fashioned, pillaging, "rape, waste, and run" extreme. In 1986, the Summitville Consolidated Mining Company started an open-pit mining operation, using a cyanide heap-leaching process to recover gold and silver from the Summitville District. This district had been mined off and on since the 1870s, with varied success; most attempts were generally not overly profitable. Located high (11,500 feet) in the eastern San Juan Mountains, with long, snowy, cold winters and potential snow slides, Summitville was a disaster waiting to happen, unless every precaution was taken to keep the cyanide solution contained in the heap-leaching pits.

Right from the start, problems plagued the project. Pollutants from the drain system were detected in nearby streams as early as the first summer of operation. The next year, again during the summer, nine cyanide spills occurred and other reckless practices were found to be taking place in the mining operation. Sadly slow on the uptake, both state and federal regulatory agencies eventually started investigating illegal discharges into streams and other violations. Their foot-dragging, plus general bureaucratic inefficiency, tragically allowed the situation to get worse. From May 1, 1989, through June 20, 1991, mine water laced with copper, zinc, and cyanide was released, or leaked, into nearby creeks and eventually threatened the Rio Grande in the downstream San Luis Valley. Because of a bad design, snow took out part of the pit where the ore and cyanide pile percolated and allowed cyanide to escape, a fact that only exacerbated the contamination problem. The acidic leakage threatened ranching, farming, and the communities in the vicinity of the mine.

Then, on December 4, 1992, the company and its parent (Galactic Resources Inc. of Vancouver, Canada) filed for bankruptcy protection. By that time, it faced $40 million in environmental stabilization costs. Sneaking out at midnight on December 15, the irresponsible, "not give a damn," arrogant company walked off and abandoned the site, leaving the cyanide and other toxic material to percolate through the now-deserted operation. An estimated 150 or so million gallons of spent cyanide processing solutions remained in the heap. The EPA took over the site the next day and declared it a Superfund cleanup site. It was discovered that documents had been falsified about environment matters, among other things. The recriminations started immediately. Who was to blame? Why had this been allowed? How could it be cleaned up?

In the end, the American taxpayer rather than the erring company paid for the cleanup, which by 1998 had cost $120 million, with another estimated $50 million more to come, before the mess was resolved. Hearings were held, fines set, plea agreements reached, and company personnel placed on probation. The company pleaded guilty to forty felony counts, the majority related to the discharge of unauthorized pollutants. According to Ken Fimberg, the assistant U.S. attorney involved in the case, the result "sets a tone that significant crimes were committed at the mine." As the *Denver Post* (June 16, 1995) correctly pointed out, "Summitville came to symbolize what environmentalists saw as everything wrong with the mining industry and the laws that govern it. It became a rallying point for efforts to reform mining laws." The fallout resulted in Colorado's mine-permitting laws being changed and strengthened, but yet again federal mining reform efforts failed.[6]

The EPA received a share of the blame as well. As a *Denver Post* editorial (February 3, 1996) bluntly stated, "there's a worrisome lack of understanding in Washington about the need for strong, but efficient, environmental protection." In the same editorial, the state received its share of blame for not monitoring the gold operations, because it "didn't have the staff, money or political will to deal with the emergency." Although the Colorado mining industry in general was not involved in this awful episode, in the public's mind it stood in the dock, guilty as charged. It would take years to overcome the Summitville black eye; meanwhile, opponents of mining had gained a wonderful weapon with which to club the industry.

Cyanide continued to be in the public eye because of problems elsewhere in the United States. In other places in the world with less environmental concern and regulation, the problem often was worse. An attempt to ban the use of cyanide to extract gold died in a Colorado senate committee in January 2003. Both sides marshaled the familiar arguments, and, of course, Summitville was dragged into the debate. One side saw problems to come "once mines closed"; the other "felt the problem did not exist." The debate drags on, awaiting another day or another accident.

Colorado mining did much better in other environmental arenas. A pollution problem (both drainage and blowing dust) existed on Red Mountain, which led to Superfund litigation. It took nine years to resolve this issue; parties included the state, San Juan and San Miguel Counties, and the Newmont Gold Company that owned the Idarado mine. In complete contrast to what had happened at Summitville, and after a bit of wrangling, an agreement was reached and reclamation was started. It included stabilization of tailings piles, revegetation, remediation of mine portals, and, interestingly, a blood lead screening program at Telluride for local children under the age of 72 months.

Silverton, too, had environmental problems after Standard Metals closed, but instead of declaring war, it worked them out with the company. While the EPA considered making the Silverton watershed a Superfund site, the newly created Upper Animas River Stakeholders Group charted their own way to a better tomorrow. The group surveyed other mines that were contributing to drainage pollution and set about cleaning up the problem. Its work included covering shafts, installing drain pipes, planting grass on waste areas, removing or covering tailing piles, and even stabilizing old buildings to preserve the area's proud mining heritage. The government helped on this project through its Abandoned Mine Lands program. The 145-square-mile Animas watershed was "one of several chosen in 1997 to be a national pilot project for mine remediation." As the *Denver Post* (October 7, 2007) observed, "For the past decade, government agencies and organizations along with a band of Silverton residents have worked together to clean up and close up—but at the same time save—San Juan County mines."

The 1983 EPA decision to clean up Leadville, another Superfund site, did not sit so well with locals. Summarizing their position clearly was a T-shirt saying "Shove it up your Yak," referring to the problem of the Yak Tunnel, which oozed drainage into California Gulch. Said one critic, "this has just been an extensive landscaping project, and it is a waste of everybody's money." From the EPA's perspective, more important were the facts that acid mine runoff contaminated the Arkansas River, tailings piles leached heavy metals, and "some of the town's children" had "unhealthy levels of lead in their blood." Local criticism of the cost, scope, and need for the project did not deter the agency. It removed and stabilized tailings piles and removed contaminated soil from yards to "reduce lead exposure in children." The agency also leveled and re-landscaped mining sites, built a water treatment plant, and did other work that many locals felt was destroying their mining heritage. Twenty-four years and $150 million later, the cleanup was still not complete.[7]

Colorado has a host of similar problems throughout its mining regions, where abandoned mine shafts, slag heaps, and portals drain mineralized water into nearby streams, polluting the local and downstream environment. Nor were mines the only activity that left behind environmental problems. For example, in Colorado Springs, where residential space brings premium prices, a housing subdivision was built on the Golden Cycle mill's old tailings pile. While it was decided that "radon infiltration" posed no threat, people were warned against putting in backyard vegetable gardens or children's play-grounds on such "soil." Dumps clutter and mar the landscape, and old mines present a danger to all who venture underground, whether from curiosity or in a search for souvenirs. State and local groups and agencies have become increasingly diligent in closing shafts and portals, but much still remains to be accomplished as the twenty-first century moves along. Abandoned mines have no legally reachable owners, so the public is left to clean up "after the ball is over." The problem remains so huge, with the number of abandoned mines scattered out there in the mountains, that no solution will be easy or economical.

The popular expression "What comes around, goes around" is certainly true of mining. Scams were not confined to isolated occurrences back in the early days. That was abundantly shown in the activities of the Dixilyn Mining Corporation, controlled by a Texas oil company, in its "wildcat" adventures with the Old Hundred Mine. The property had been mined on and off since the 1870s, with little success in the twentieth century. In 1967, Dixilyn leased the property and started diamond-drilling and driving tunnels, drifting, and working in the mine. It acquired the Pride of the West Mill at Howardsville, leased other mines, and pushed ahead with all the projects. The company encountered plenty of water and lots of quartz, but little or no gold or silver. However, announcements in 1969 claimed to have found high-grade veins. Stock values jumped, investors came to see, and the company touted its up-and-coming property wherever and whenever it had the opportunity.

A year later, the wonderful promises and ore values still had not been substantiated. High-grade gold seemed to elude Dixilyn, no matter where it drilled or mucked. They had plenty of zinc, though—enough that by late spring of 1971, neighboring Standard Metals was talking about a lawsuit, as it appeared that Dixilyn was mining on Standard's property. The issue was ultimately settled out of court, with Dixilyn paying for the ore. In the meantime, even as layoffs occurred and rumors began to circulate, the company steadfastly maintained a stiff upper lip about prospects for the future. The finale came that fall, when the last employees were laid off, the mill was shut down, and Dixilyn wrote "off the extraordinary loss of $6 million." The year 1975 saw the bitter end: a foreclosure sale for failure to pay the company's debts and the demolition of the company's buildings at the base of Cunningham Gulch.

"Standard Metals had the muck, Dixilyn had the money" was a saying around Silverton a generation ago. In truth, that might be the epitaph for many Colorado mining ventures over the past generations—money and high-grade ore seemed unable to get together. The industry left behind broken dreams, shattered hopes, financial ruin, and the physical remains of many a mining venture.

As gold and silver mining declined, though, its heritage started to produce revenue in a different way. Mines, mining sites, ghost towns, and declining mining communities have fascinated tourists for generations. In Colorado's post-World War II years, all these things gradually emerged as a major tourist attraction when four-wheel-drive vehicles allowed easy access to once-isolated spots.

By the late twentieth century, and on into the twenty-first, mining had become a tourist mecca. Mine tours could be taken in Silverton, Ouray, Cripple Creek, Breckenridge, and Idaho Springs, for example. Silverton and

Idaho Springs both had old mills that visitors could tour. Because Silverton was fortunate enough to have all the major factors needed for successful mining, visitors get a more complete picture. Tourists can ride the narrow-gauge train, the Durango & Silverton, just as so many did a hundred or more years before. They can tour the Old Hundred Mine (which is more profitable now than when Dixilyn mucked out of it), and explore the completely equipped Standard Metals Mill, which was donated to the San Juan County Historical Society. A walk through Silverton lets tourists see architectural styles of homes, city and county buildings, and commercial structures. A quick change of weather could give the visitor a feel for what it was like to live in a high, isolated mountain valley. Maybe some lucky folk might even find a souvenir bit of gold ore as they "prospected" in the mountains—hopefully not taken from someone's patented claim!

Mining might be only a pale shadow of what it once was and what it once meant to Colorado, but it is still in evidence in a variety of venues. Two other factors also carried some of its legacy forward as the twentieth century ended: gambling and historic preservation. In 1990, Colorado voters approved limited-stakes gambling for three declining historic mining towns, Central City, Black Hawk, and Cripple Creek. Within a year, Colorado had hit a gambling bonanza. By law, tax monies generated from gaming were to be distributed to the towns and counties that hosted the casinos, to the state historical preservation fund, and to the Colorado general fund.

Black Hawk, the nearest to Denver and its suburbs, profited the most, but at the cost of losing its historic character (during the gambling renovations, many "historic" buildings were moved to a designated spot). Central City and Cripple Creek did not enjoy quite the same influx of prosperity, but local historic preservation gained immensely. Preservation projects throughout the state, from prairie farming towns to foothill towns, mountain mining camps, and Western Slope ranching communities, received a monetary shot in the arm. Those funds especially helped depressed mining towns to preserve the ambiance and structural fabric of their earlier age. The preservation and renovation of historic buildings in such communities give the visitor a clearer appreciation and understanding of the urbanization that was brought about by mining and the life and times of another age.

Leadville shows the entire evolution of mining, from the 1870s into the twenty-first century, and the National Mining Museum and Hall of Fame there displays it all and more. Georgetown, Ouray, Central City, Creede, Rico, and Breckenridge likewise preserve their own mining heritage, as do many other communities throughout the Colorado Rockies. Aspen and Telluride have maintained some of their mining past, but have turned most of their attention to becoming resort destinations for skiers and the rich and famous. A few buildings have been preserved in sites such as Ashcroft, Animas Forks, Silver Cliff, and Nevada City. Fairplay's reconstructed South Park City, which features numerous buildings salvaged from other sites, gives visitors the feel of a reasonably authentic mining community. Historical societies, museums, historical markers, and individual buildings also recapture and preserve mining's heritage in a variety of ways. It is all beckoning for those with interest and the time to savor a taste of the past.

Photographic Essay

Colorado Mining in the Twentieth and Twenty-First Centuries

By the turn of the century into 1900, it had become obvious that mining would not play the premier role it once had within the state's economy. Much of the romance, excitement, and newness of the "rush" days had disappeared, replaced by an industrial America of union/management conflicts and, elsewhere, of small mines struggling to survive.

In the decades of the twentieth century, the role of mining diminished steadily, and most of the once-flourishing mining communities became more of a tourist attraction than an economic pillar of the state economy. Even at twenty-first-century Cripple Creek, where modern mining methods extracted more gold than ever, mining drew little attention. Often it seemed that there was more interest in preserving the past than in planning the present and future of mining.

Still, the photographers of the twentieth century left behind a record of an industry in transition and decline. Commenting about the South African mining excitement of 1897, that former miner Mark Twain observed, "I had been a gold miner myself, in my day, and knew substantially everything that those people knew about it, except how to make money at it." With few exceptions, that is the epitaph of Colorado mining after World War I, a story that photography documented.

Bullion Tunnel, the main access to the Smuggler Union Mine, high above Telluride. The Tomboy Mine was in the basin to the right.

COURTESY OF MARK & KAREN VENDL.

Miners in Cripple Creek's Lillie Mine, 900-foot level.

COURTESY OF MARK & KAREN VENDL.

Mass labor meeting during the Cripple Creek labor troubles of 1903–1904.

COURTESY OF MARK & KAREN VENDL.

The ubiquitous blacksmith shop was a fixture at mines large and small.

COURTESY OF COLORADO HISTORICAL SOCIETY.

A superintendent could travel around his mine fast and easily on his special bicycle. This one is at the Newhouse Tunnel next to Idaho Springs.

COURTESY OF SPECIAL COLLECTIONS, COLORADO COLLEGE.

The Timberline Dredging Company operated a dredge near Fairplay in 1939, leaving behind its tailings pile.

COURTESY OF U.S. FOREST SERVICE.

Horses pull and push to get this boiler delivered to its destination, somewhere west of Silver Plume.

COURTESY OF CHRIS BRADLEY.

Even in the 1920s, mule and burro power still pulled ore carts and hauled ore to the railroad or smelter.

COURTESY OF CHRIS BRADLEY.

Double- and single-jack drilling contests were a popular attraction well into the twentieth century, and are still held in the twenty-first.

COURTESY OF DENVER PUBLIC LIBRARY, WESTERN HISTORY COLLECTION.

Locomotives replaced animals in many Colorado mines by the 1920s. This one is pulling a car used to transport miners.

COURTESY OF CHRIS BRADLEY.

During the hard times of the 1930s depression, Uncle Sam sponsored classes that taught how to pan for gold. How successful were these women? Their results are lost to history.

COURTESY OF DENVER PUBLIC LIBRARY, WESTERN HISTORY COLLECTION.

Snow has always posed a threat to Colorado mining. This snow slide hit a Telluride mill.
COURTESY OF DUANE A. SMITH.

Miners drifting with jackleg drills at the Bulldog mine, near Creede.
COURTESY OF HOMESTAKE MINING COMPANY.

Trucks haul thousands of tons of ore per day at the Cripple Creek & Victor open-pit mine.

COURTESY OF CRIPPLE CREEK & VICTOR GOLD MINING COMPANY.

For a generation, women have been involved in all phases of mining.

COURTESY OF CRIPPLE CREEK & VICTOR GOLD MINING COMPANY.

The heritage of silver and gold mining can be seen throughout the mountains of Colorado. This shaft house sits in the Red Mountain district.

COURTESY OF JOHN NINNEMANN.

Epilogue: A Tale Well Told

David Lavender, in his *One Man's West,* tells the tale of a heaven-bound prospector/miner stopped at the pearly gates by St. Peter. Inquiring why, the miner receives this startling answer:

> "You can't come in. I passed a bunch of miners last month, and now they're assaying their harps, digging up the golden streets and stoping out the Elysian fields. It's driving the rest of the angels frantic."
>
> "I'll make a bargain with ye, Pete," the prospector said. "If ye'll let me in I'll get rid of every mother's son of them hard-rock stiffs."
>
> The two struck a bargain, the gates opened, and in went the prospector. He quickly started a rumor . . . that there was a big gold strike in the vortex of hell. In an instant the entire mining population cast aside their crowns and disappeared over the rim. As the last one passed from sight, the old prospector suddenly grabbed up a pick and set out in pursuit. Amazed, St. Peter watched him start off.
>
> "So long, Pete!" he called over his shoulder. "I'm going too!"

This classic yarn, nearly as old as Colorado mining and extant in several variations, neatly captures the allure—the "fever"—of mining.

How else may what transpired, in the years between 1859 and 2009, be comprehended, appreciated, and perhaps even logically explained? Why would men and women endure the uncertain life they did, starting with the Pike's Peak rush, were it not for the lethal attractions of silver and gold? This has been their story, an industry's saga, and Colorado's legacy.

Over a century and a half, the mining industry's contribution to the state, and to generations of Coloradans, has been virtually infinite. Starting with 1859, those include lure, settlement, territorial status, statehood, publicity, development, investment, diversification of the economy, employment both within and outside the industry, transportation, tourism, advances in mining and smelting technology, and urbanization, to name just a few of the obvious influences and consequences. One cannot measure the impact on the lives of the hundreds of thousands of people who lived, worked, invested in, and visited Colorado mines and mining communities, or the influence of mining's legacy on the tourists who still visit this fabled land.

The stories and personalities of the participants colored much of Colorado's history for its first half-century. The overwhelming majority of its political, business, and social movers and shakers had ties to mining, and their influence lasted well past the turn of the twentieth century. Though they were well known, equally important to the story are the vast multitudes of people who passed through camps, towns, and mining occupations but left little trace of their having come and gone. They were the heart and soul of an industry and their generation.

The poet Thomas Hornsby Ferril captured their lives in his poem "Magenta." The poem's narrator, looking over a declining Central City, told of its lost glory and the women and men who resided there:

> The town was high and lonely in the mountains;
> There was nothing to listen to but the wasting of
> The glaciers and a wind that had no trees.
> And many houses were gone, only masonry
> Of stone foundations tilting over the canyon,
> Like hanging gardens where successful rhubarb
> Had crossed the kitchen sill and entered the parlor.
>
> . . .
>
> The men would measure in cords the gold they hoped
> To find, but the women reckoned by calendars

Of double chins and crow's feet at the corners
Of their eyes.

Their story tells the real saga of Colorado mining during its glory years.

What magnificent years those were, even if one counts only the ore mined out of the river bottoms and mountain depths. The industry was the major economic pillar of the state until World War I. In the decades from 1859 through 1922, gold and silver mining produced more than $1,163,829,926. In the years that followed, production never equaled that level. In the past decade, though, thanks to increasing world prices and Cripple Creek's continuing production of more than 200,000 ounces per year, production—in dollars' worth, at least—rivals that of earlier years.

What this meant in 2009, with the average spot price for gold at $926 on June 15, was that production topped $200 million. Nothing demonstrates the evolution of Colorado mining better than Cripple Creek's open-pit operation, where gargantuan trucks move mega-tons of ore daily. This is the twenty-first-century version of "gold fever," which, in many respects, Coloradans have never outgrown. However, the volume of gold ore "dug" in one mine in one day would have startled old-timers. Of course, the price of gold today would also amaze the miners of yesterday. Even their wildest dreams would not have included a per-ounce price hovering around $900; in mid-March of 2008, an ounce of gold topped $1,000 for a while before slipping down again. Both price levels would have been unbelievable only a decade earlier.

A *Denver Post* editorial (March 15, 2008), "This Isn't Your Father's Gold Rush," returned to another theme that had been around for a generation: "an overhaul of the [1872] mining law is long overdue." It proposed that royalties from "gross income" be, among other things, tied to a mine cleanup fund. A "Hardrock Mining and Reclamation Act" had already passed the House and awaited Senate action. Still, Congress has been down this road before, and the end is still not in sight in mid-2009.

The miner of 1859, 1879, and 1909 would be amazed at the continuing evolution in the mining industry, and Colorado's development, since he dug his lonely way into the mountainside. One item that would certainly catch his attention would be the wages paid his industrial descendants—but he would need to take a good look at the cost of living today before he succumbed to envy. He would look about and see only large mining corporations. The wandering jackass prospector is long gone, along with most of the small-time, independent operators. Meanwhile, Uncle Sam watches over every miner's shoulder. That old-timer would, no doubt, be surprised at today's safety measures, today's equipment, and today's mining methods, not to mention the cost of that mining equipment. He would probably bemoan the loss of the independence and freedom that were once hallmarks of the industry, but which are now only part of the legend of the "rush" days.

Nothing might shock him more than the environmental revolution of the past generation. Nor would the harsh criticism of mining and its legacy, during the 1960s and 1970s, be appreciated or understood. Criticism continues into the twenty-first century, though generally with less emotion and public outcry unless there is an environmental accident. Some of the criticism has certainly been warranted; some has not. Even today, in some respects, most non-mining Americans do not understand the significance of mining to the present and the future. Once again, that bumper sticker reminds all who see it: "If it can't be Grown, it has to be Mined." Coloradans face the multidimensional problem of escaping their mining past, accepting that past, and coming to grips with the legacy of that past. Then they must come to grips with the future.

Of the four most important Colorado mining districts (Central City, Leadville, San Juans, and Cripple Creek), only the last is still actively mined today on a major scale. The others mine tourists' pocketbooks, thanks to their historical legacies: They profit from their past by recalling their days of fame and fortune. In many senses, they, and their counterparts in other states, have made legends out of the things they want to believe. In some ways, mining provides a more constant source of income for the old mining camps and towns in the twenty-first century than it ever did when it underwrote their settlement and development. Thus, mining continues to play a role throughout the mountains and valleys, with fact and reality so often intermixed with legend and folklore that the two are sometimes hard to separate.

Colorado mining, which gave the towns their mines, enabled their birth, furnished their sustenance, nourished their growth, and delimited their lives, "is dead my friend," in the words of that quintessential Westerner Charlie Russell, "but writers hold the seed and what they saw will live and grow again to those who read." Horace Tabor, the epitome of Colorado mining's reality and legend, apparently understood this. He placed these words of the poet Charles Kingsley on the drop curtain of his splendid architectural achievement, the Tabor Grand Opera House:

> So fleet the works of men back to the earth again;
> Ancient and holy things fade like a dream.

Russell's prediction stands against the fact that neither fiction writers nor Hollywood ever took much of a fancy to mining. The legend of the cowboy on a horse, riding the open range, driving a herd along a trail, and galloping into town for fun and frolic somehow seems more romantic, more exciting, and more typically Western to the public. A miner working an underground shift in a dangerous, dark environment, or a millhand laboring to separate gold and silver from mere rock, does not attract as many readers or viewers as stories of the wide-open spaces.

Occasionally a mining town gains some literary, television, or film attention, but it usually comes as a function of the urban environment rather than the mining that was the industrial base of its existence—or it becomes simply the backdrop for a western morality play. The musical, *The Unsinkable Molly Brown*, is a good example. Although Molly had strong ties to Leadville mining, her primary fame sprang from her survival of the sinking of the *Titanic.*

Perhaps it is only fitting that grand opera—a genre known for its larger-than-life characters, often-melodramatic plots, and towering passions—best recaptures the Leadville scene, the late nineteenth-century Colorado era, and mining's highs and lows. *The Ballad of Baby Doe* ardently and sensitively does what historians and writers have had difficulty achieving: It elicits a spirit, an essence, an emotion that opens a door to the past and brings it all back to life.

The Tabor story has fascinated for generations, as no other story from those mining years has ever done—possibly because Horace Tabor's tale has it all: hard times, sacrifice, sudden success, loyalty and betrayal, politics, scandal, heartbreak, and financial ruin. Though fictionalized, of course, John Latouche's solidly fact-based libretto, along with Douglas Moore's splendid music, capture and present the truth better than any mere written account ever could. On wings of emotion and song, the opera reveals the heart, the soul, the ambiance, and the frame of mind of an era and a generation.

The Ballad of Baby Doe begins in November 1879, on the opening night of Tabor's opera house—Leadville's symbol of culture and its grand aspirations. In that flush year of Leadville's fame and fortune, the suddenly famous Tabor tells his cronies, who are taking a breather from the festivities, that times could not be better. Indeed, they could not: silver mining, Leadville, and Colorado had become synonymous.

I dug by day and dug by starlight.
I'm an honest son of labor.
Dug my way right through to Hell;
Satan said "Why here comes Tabor!"

Dig, dig you gophers, dig them holes,
Dig away to save your souls!
There's mountains galore of silver ore;
It's cheap in Colorado.

That evening, mining was king in Colorado, the future only held more gold and silver bonanzas, and everything and anything seemed possible. The rest of the story, mirroring the rise and fall of Colorado mining, unfolds through dizzying successes to heart-wrenching failures.

In the opera's last scene, Tabor—old, weary, semi-delirious, and nearly bankrupt—wanders onto the stage of his empty opera house. He looks about and finds himself reliving his life:

Wow! He's struck it!
Struck the Little Pittsburg—ten thousand a week!
Struck the Chrysolite—twenty thousand a week!
Struck the New Discovery,
The Dives and the Winnemucca—Count up the millions!

Recalling those excitements and triumphs, Tabor realizes that they all took place many yesterdays ago, far distant though only twenty years past. The phantom of his first wife, Augusta, then confronts Horace, sharply rebukes him, and harshly forces him to face the issue of his career and significance: "You are going to die, Horace Tabor, and you die a failure." Tabor responds with an aria that might easily echo through the decades as an epitaph for what once was and never would be again:

How can a man measure himself?
The land was growing, and I grew with it.

How can one measure Tabor, his predecessors and contemporaries, and the mining generations since? One thing is clear: They did not fail. The land has grown remarkably within the 150 years of this story and its saga has taken on epic proportions. Colorado gold and silver mining played a major role in that saga, and even if, in the twenty-first century, that "gold and silver echo" that Coloradans once heard "roar in the wind" no longer resounds around the mountains and down the canyons, its story is not yet finished.

Notes

CHAPTER 1: PIKE'S PEAK OR BUST

1. Quoted in William Byers, *Hand Book to the Gold Fields of Nebraska and Kansas* (Chicago: D. B. Cooke, 1859), 9.

2. *Kansas Weekly Herald* (Leavenworth), July 24, 1858.

3. *Journal of Commerce* (Kansas City), July 30, 1854.

4. Cited in LeRoy Hafen, *Pike's Peak Gold Rush Guidebooks of 1859* (Glendale, Cal.: Arthur H. Clark, 1941), 76.

5. Byers, *Hand Book,* 113. See also LeRoy Hafen, *Overland Routes to the Gold Fields, 1859* (Glendale, Cal.: Arthur H. Clark, 1942). Most of the reports and quotes in this section are from these two books.

6. Quoted in Elliott West, *Contested Plains: Indians, Goldseekers, and the Rush to Colorado* (Lawrence: University of Kansas Press, 1998), 127–128.

7. Quoted in Hafen, *Overland Routes*, 280–281.

8. Quoted in Hafen, *Overland Routes*, 47–48, 156, 260–261, 276, 314.

9. *Missouri Democrat,* May 31, 1859, quoted in Hafen, *Overland Routes*, 315.

10. Thomas M. Marshall, *Early Records of Gilpin County, Colorado 1859–1861* (Boulder: University of Colorado Press, 1920), 1–4.

11. Horace Greeley, *An Overland Journey from New York to San Francisco in the Summer of 1859* (New York: Alfred A. Knopf, 1964), 98–106.

12. Greeley, *Overland Journey*, 127–129.

CHAPTER 2: 1859: THE YEAR DREAMS BECAME REALITY

1. Thomas Marshall, ed., *Early Records of Gilpin County, Colorado 1859–1861* (Boulder: University of Colorado, 1920), 10–12.

2. *Tom* is a shortened version of *long tom*, a sluice box that stretched a hundred feet or more. *Rockers* are contraptions that looked somewhat like a child's cradle but with a sieve-like tray at the top of one end, several wooden cleats across the length of the tray, and an opening at the opposite end.

3. Marshall, *Early Records*, 12–15, 55, 66, 114, 125, 172–173, 185, 195, 202, 228, 242, 246, 258.

4. Marshall, *Early Records*, 12–13, 15, 74, 151, 192, 244, 256.

5. L. D. Crandall to nieces, July 17, 1859 (Madison: State Historical Society of Wisconsin).

6. *Rocky Mountain News*, June 18, 1859. Unless otherwise cited, all newspaper references in this chapter are from the *News*.

7. Frank Fossett, *Colorado: Its Gold and Silver Mines* (New York: C. G. Crawford, 1879), 122; Ovando J. Hollister, *The Mines of Colorado* (Springfield, Mass.: Samuel Bowles, 1867), 68, 72.

8. Copies of the Tabor interviews are found in the Bancroft Library and the Colorado State Historical Society.

9. Henry Villard, *Past and Present* (Princeton, N.J.: Princeton University Press, 1932 reprint), 57; Robert H. Shikes, *Rocky Mountain Medicine* (Boulder, Colo.: Johnson, 1986), 26, 30–31, 39–40. See also Ronald C. Brown and Duane A. Smith, *No One Ailing Except a Physician* (Boulder: University Press of Colorado, 2001).

CHAPTER 3: 1860–1864: "TO EVERYTHING THERE IS A SEASON"

1. William A. Crawford, to Mina, June 17, 1860 (Madison: State Historical Society of Wisconsin).

2. Ovando J. Hollister, *The Mines of Colorado* (Springfield, Mass.: Samuel Bowles, 1867), 106–107.

3. Crawford to Mina, June 17, 1860; September 2, March 31, and October 13, 1861; May 11, 1862 (Madison: State Historical Society of Wisconsin).

4. Hollister, *Mines of Colorado*, 108–109.

5. Hollister, *Mines of Colorado,* 111–112.

6. Sources for the section on the California Gulch rush: *Weekly Rocky Mountain News*, March 28, July 25, and August 15, 22, 1860; May 8, 15, and August 29, 1861; Augusta Tabor, "Cabin Life," Bancroft Library, 2–3; Don and Jean Griswold, *History of Leadville and Lake County, Colorado* (Denver: Colorado Historical Society, 1996), vol. 1, 43–46; Alice Polk Hill, *Tales of Colorado Pioneers* (Denver: Pierson & Gardner, 1884), 277.

7. Hollister, *Mines of Colorado*, 122; Samuel Mallory, letter, July 8, 1860, *Colorado Magazine* (May 1931), 114.

8. *Rocky Mountain News*, July 18, 1860; June 5, 1861; and July 16, 1863; *Miners' Record*, August 17 and September 14, 1861; *The Rocky Mountain Directory and Colorado Gazetteer for 1871* (Denver: S. S. Wallihan & Co., 1871), 169–171.

9. *Rocky Mountain News*, April 26, 1862; *Weekly Rocky Mountain News*, September 11, 1862 and January 8, 1863; *Colorado Republican*, June 26, 1862; *Colorado Mining Life*, September 13, 1863.

10. Samuel Leach to George Leach, October 25, 1862, *The Trail* (March 1926), 11. For Pat Casey, see Duane A. Smith, *The Birth of Colorado* (Norman: University of Oklahoma Press, 1989).

11. Maurice O'Connor Morris, *Rambles in the Rocky Mountains* (London: Smith, Elder, 1864), 124; *Weekly Rocky Mountain News*, July 16, 1863.

12. See Smith, *The Birth of Colorado*, 97–99, 159–163, 195–197.

CHAPTER 4: 1864–1869: "GOOD TIMES A-COMIN"—SOMEDAY

1. James W. Taylor, "The Mineral Resources of the United States East of the Rocky Mountains," in *Reports on the Mineral Resources of the United States* (Washington, D.C.: Government Printing Office, 1868), 9.

2. See chapters 1–3 in Liston E. Leyendecker, Christine Bradley, and Duane A. Smith, *The Rise of the Silver Queen* (Boulder: University Press of Colorado, 2005). See also issues of the *American Mining Journal*, 1867–1869.

3. Rossiter Raymond, *Statistics of Mines and Mining* (Washington, D.C.: Government Printing Office, 1870), 368–369.

4. Ovando J. Hollister, *The Mines of Colorado* (Springfield, Mass.: Samuel Bowles, 1868), 318.

5. James W. Taylor, "Gold Mines East of the Rocky Mountains," in *Reports upon the Mineral Resources of the United States* (Washington, D.C.: Government Printing Office, 1867), 328; Taylor, "Mineral Resources," 8.

6. Taylor, "Gold Mines," 327; Raymond, *Mines and Mining*, 347–348.

7. Raymond, *Mines and Mining*, 362.

8. Nathaniel Hill to "My Dear Wife," June 23, 30, August 8, 11, September 4, 22, and October 3, 16, 1864 (Hill Papers in Western History Collection, Denver Public Library). For a detailed examination of Hill and his smelter, see James E. Fell Jr., *Ores to Metals* (Lincoln: University of Nebraska Press, 1979), 11–54.

9. Raymond, *Mineral Resources,* 348–349 and 356–362.

10. Taylor, "Gold Mines," in *Mineral Resources* (1867), 351–355; Rodman Paul, *Mining Frontiers of the Far West* (New York: Holt, Rinehart & Winston, 1963), 169–173.

11. Raymond, *Mines and Mining,* 342–343.

CHAPTER 5: 1870–1874: BONANZA! "THREE CHEERS AND A TIGER"

1. Rossiter Raymond, *Statistics of Mines and Mining* (Washington, D.C.: Government Printing Office, 1872), 287.

2. James D. Hague, "Mining Engineering and Mining Law," *Engineering and Mining Journal* (October 20, 1904), 6, 8.

3. Rossiter Raymond, *Statistics of Mines and Mining* (Washington, D.C.: Government Printing Office, 1874), 287. See also Rossiter Raymond, *Statistics of Mines and Mining* (Washington, D.C., 1870), 364.

4. Rodman Paul, *Mining Frontiers of the Far West* (New York: Holt, Rinehart & Winston, 1963), 173–174; Rossiter Raymond, *Statistics of Mines and Mining* (Washington, D.C.: Government Printing Office, 1873), 453–454, 456. The entire law is printed in this section.

5. *Mining and Scientific Press,* October 11, November 15, 22, and December 13, 1873.

6. Raymond, *Statistics of Mines and Mining* (1870), 293.

7. Thomas T. Read, *The Development of Mineral Industry Education in the United States* (New York: American Institute of Mining and Metallurgical Engineers, 1941), 88–90.

8. Herbert Hoover, *The Memories of Herbert Hoover* (New York: McGraw-Hill, 1901), vol. I, 131; Clark C. Spence, *Mining Engineers and The American West* (New Haven: Yale University Press, 1970), 370; Raymond, *Statistics of Mines and Mining* (1873), in ch. XIX, "American Schools of Mining and Metallurgy."

9. Raymond, *Statistics of Mines and Mining* (1874), 386.

10. *Summering in Colorado* (Denver: Richards, 1874), 21, 24; Rossiter W. Raymond, *Camp and Cabin* (New York: Fords, Howard & Hulbert, 1880), 235; Sidney Glazer, ed., "A Michigan Correspondent in Colorado, 1878," *Colorado Magazine* (July 1960), 215; *Engineering and Mining Journal,* September 7, 1877, 50.

11. *Summering in Colorado,* 21–22; James Rusling, *Across America* (New York: Sheldon, 1874), 66.

12. Joseph T. Gordon & Judith A. Pickle, eds., *Helen Hunt Jackson's Colorado* (Colorado Springs: Colorado College, 1989), 18.

CHAPTER 6: 1875–1880: "ALL ROADS LEAD TO LEADVILLE"

1. Rodman W. Paul, ed., *A Victorian Gentlewoman in the Far West: The Reminiscences of Mary Hallock Foote* (San Marino, Cal.: Huntington Library, 1972), 172.

2. Rossiter W. Raymond, *Statistics of Mines and Mining* (Washington, D.C.: Government Printing Office, 1877), 284–286.

3. James E. Fell, Jr., *Ores to Metals* (Lincoln: University of Nebraska Press, 1979), 44–46, 137–138, 160–162; Rossiter Raymond, *Statistics of Mines and Mining* (Washington, D.C.: Government Printing Office, 1875), 379–385.

4. Charles Harvey to family, May 22, 1879 (Charles H. Harvey, Library of Congress, Washington, D.C.).

5. *Proceedings of the Constitutional Convention . . . For the State of Colorado* (Denver: Smith-Brooks, 1907), 688, 700.

6. Composite picture of Oro City taken from: *Rocky Mountain News*, August 3, 1870, October 25, 1871, July 21, 1872, and June 29 and August 31, 1873; *Engineering and Mining Journal*, February 19 and May 6, 1876, March 31, 1877; *New York Times*, May 20, 1878.

7. Henry Wood, *Territorial Assay Book*, entries for 1873–1874 (Henry Wood Collection, Huntington Library, San Marino, California); *Rocky Mountain News*, July 1872 and November 29, 1873; *Engineering and Mining Journal*, December 30, 1876.

8. Frank Fossett, *Colorado: Its Gold and Silver Mines* (New York: C. G. Crawford, 1879), 411. See also Don Griswold and Jean Griswold, *History of Leadville and Lake County, Colorado* (Denver: Colorado Historical Society, 1996), vol. 1, 134–147; New Blair, *Leadville* (Boulder, Colo.: Pruett Publishing, 1980), 27–31.

9. Fossett, *Colorado: Its Gold and Silver Mines*, 416.

10. M. H. Foote to "Helena," Fall 1880, and to "Beloved Girl," July 10, 1880 (James D. Hague Collection, Henry E. Huntington Library, San Marino, California). The previous quotations come from Foote letters of May 12, May 28, July 8, September 8, and November (no day) 1879, all from the James D. Hague Collection.

11. M. H. Foote to "James Hague," May 27, 1887 (James D. Hague Collection, Henry E. Huntington Library, San Marino, California); Mary Hallock Foote, *The Led-Horse Claim: A Romance of a Mining Camp* (Ridgewood, N. J.: Gregg Press, 1968 reprint), 11.

12. Rodman Paul, "Colorado as a Pioneer of Science in the Mining West," *Mississippi Valley Historical Review* (June, 1960), 46–47.

13. Samuel F. Emmons, "Lead Smelting at Leadville, Colorado," *Statistics and Technology of the Precious Metals* (Washington, D.C.: Government Printing Office, 1885), 295.

CHAPTER 7: THE SILVER EIGHTIES: THE BEST OF TIMES, THE WORST OF TIMES

1. Quoted in Charles Henderson, *Mining in Colorado* (Washington, D.C.: Government Printing Office, 1926), 124–125.

2. Duane Vandenbusche, *The Gunnison County* (Gunnison, Colo.: B&B Printers, 1980), chs. 10, 12, 13; Helen Hunt Jackson, "Oh-Be-Joyful Creek and Poverty Gulch," *Atlantic Monthly* (December, 1883), 755–757.

3. William Weston, *Descriptive Pamphlet . . . Silver Region* (Montrose, Colo.: Western Reflections, 2006 [reprint]), 33–35; *Rocky Mountain News*, April 17, 1883, September 1, 1885; *Red Mountain Pilot*, June 2, 1883; *Engineering and Mining Journal*, August 4, September 22, 1888, and August 3, 1889.

4. Ernest Ingersoll, "Silver San Juan," *Harper's New Monthly Magazine* (April, 1882), 688–690, 703–704; *Engineering and Mining Journal*, February 28, 1880, February 4, 1882, October 19, 1887; *The San Juan* (Silverton), May 19, 1887.

5. *Engineering and Mining Journal*, May 5, 1888. Malcolm J. Rohrbough discusses this case in detail in *Aspen: The History of a Silver-Mining Town, 1879–1893* (New York: Oxford University Press, 1986), 96–107.

6. *Congressional Record* (March 21, 1882), 2100; Henry Teller, speech on gold and silver coinage, *Congressional Record* (January 19, 1886), 30.

7. Letter to *New York Tribune*, n.d., Tabor Scrapbook 5, Colorado Historical Society, Denver; speech, January 30, 1890, Tabor Box 15, Colorado Historical Society, Denver.

8. William Jennings Bryan, *The Cross of Gold* (Lincoln: University of Nebraska Press, 1996), 17, 19.

CHAPTER 8: "THERE'LL BE A HOT TIME"

1. Mary Hallock Foote, quoted in Rodman W. Paul, ed., *A Victorian Gentlewoman in the Far West* (San Marino, Cal.: Huntington Library, 1972), 175, 200.

2. Helen Hunt Jackson, "O-Be-Joyful Creek and Poverty Gulch," *Atlantic Monthly* (December 1883), 753.

3. B. C. Keeler, *Leadville and Its Silver Mining* (Chicago: Perry Bros., 1879), 3–4.

4. Isabella L. Bird, *A Lady's Life in the Rocky Mountains* (Norman: University of Oklahoma Press, 1960 reprint), 190–191.

5. Ward Ordinance Books; Telluride Ordinance Books; Victor Ordinance Books; St. Elmo Ordinance Books; Liston E. Leyendecker, Christine A. Bradley, and Duane A. Smith, *The Rise of the Silver Queen* (Boulder: University Press of Colorado, 2005), 34–35.

6. Mabel Barbee Lee, *Cripple Creek Days* (New York: Doubleday, 1958), 75–77.

7. William Jackson, "Railroad Conflicts in Colorado in the Eighties," *Colorado Magazine* (June 1946), 17; *Leslie's Illustrated*, April 12 and May 17, 1879.

8. Mollie Dorsey Sanford, *Mollie* (Lincoln: University of Nebraska Press, 1959), 137, 139; Harriet Backus, *Tomboy Bride* (Boulder, Colo.: Pruett, 1970), 38–39, 110; Duane A. Smith, ed., *A Visit with the Tomboy Bride* (Montrose, Colo.: Western Reflections, 2003), 51; Lee, *Cripple Creek Days*, 21–22; Anne Ellis, *Life of an Ordinary Woman* (Lincoln: University of Nebraska Press, 1980 repr.), 29, 205.

9. Mildred Ekman, letter to author, June 18, 1974; Martha Gibbs, letter to author, August 22, 1974; author interview with Ernest Hoffman, March 2, 1973.

10. Baseball section, *Daily Central City Register*, September 6–8, 1871; *Miners' Register*, July 14, 1867; *Rocky Mountain News*, April 16, 1867; *San Miguel Examiner*, May 18,

1907. For more on the Leadville Blues, see Duane A. Smith, "Baseball Champions of Colorado: The Leadville Blues of 1882," *Journal of Sport History,* Spring 1977.

11. Rev. J. J. Gibbons, *In the San Juan* (Telluride, Colo.: n.p., 1972 repr.), 20; George Darley, *Pioneering in the San Juan* (Chicago: F. H. Revell, 1899), 21, 22, 37, 124.

CHAPTER 9: "THE EVERLASTING LOVE OF THE GAME"

1. Mabel Barbee Lee, *Cripple Creek Days* (New York: Doubleday, 1958), 270. See also David Lavender, *Red Mountain* (Ouray, Colo.: Western Reflections, 2000 [reprint]), 517.

2. Henry M. Teller, "Silver Coinage," January 6, 1892 (Washington, D.C.: Government Printing Office, 1892), 3–13; Henry M. Teller, "The Silver Question," April 20, 1892 (Washington, D.C.: Government Printing Office, 1892), 3, 28.

3. *Rocky Mountain News,* July 20, 1893; *Rocky Mountain Sun,* July 1, 1893; *Silverton Standard,* July 22, 1893; *Pagosa Springs News,* August 11, 1893.

4. Colorado Bureau of Labor Statistics, *Effects of Demonetization of Silver on the Industries of Colorado, July 1 to August 31, 1893* (Denver: Smith-Brook Printing, 1893), 2–33.

5. *Constitution and By-Laws of the Eldora Miners' Union No. 45* (Eldora, Colo.: Miner Print, 1898), 2–23.

6. William Jennings Bryan, *The Cross of Gold* (Lincoln: University of Nebraska Press, 1966), 28.

7. Vachel Lindsay, "Bryan, Bryan, Bryan, Bryan"; available at http: / / www.geocities.com / vachellindsaybryan

8. T. A. Rickard, quoted in Robert A. Trennert, *Riding the High Wire* (Boulder: University Press of Colorado, 2001), 1; see also ibid. at 66.

9. Frank Crampton, *Deep Enough* (Norman: University of Oklahoma Press, 1982 [reprint]), 22.

10. Anne Ellis, *The Life of an Ordinary Woman* (Lincoln: University of Nebraska Press, 1980 [reprint]), 206.

11. Robert Service, "The Prospector," in *The Best of Robert Service* (New York: Dodd, Mead, 1953), 56.

CHAPTER 10: 1900–1929: LOOKING FORWARD INTO YESTERDAY

1. James E. Fell, Jr., *Ores to Metals: The Rocky Mountain Smelting Industry* (Lincoln: University of Nebraska Press, 1979), 232–236.

2. George Graham Rice, *My Adventures with Your Money* (Las Vegas: Nevada Publications, 1986 [reprint]), 25.

3. Dan Plazak, *A Hole in the Ground with a Liar at the Top: Fraud and Deceit in the Golden Age of American Mining* (Salt Lake City: University of Utah Press, 2006), 201, 288.

4. *Cripple Creek Times*, August 15, 1903. See also George G. Suggs, Jr., *Colorado's War on Militant Unionism* (Detroit, Mich.: Wayne State University, 1972), chs. 5 & 7.

5. *Reports of Listed Companies* (Colorado Springs, Colo.: Colorado Springs Mining Stock Association, 1907), 1–3.

6. For a general overview of these times, see Suggs, *Colorado's War*, 84–100; Robert Guliford Taylor, *Cripple Creek* (Bloomington: Indiana University Press, 1966), 116–120.

7. Carroll H. Coberly, "Ashcroft," *Colorado Magazine*, vol. 37, 1960, 88–94.

8. Charles A. Bramble, *The A B C of Mining* (Chicago: Rand, McNally & Co., 1898), 145, 147–148.

9. *Gunnison News-Champion*, February 2, 1912; Duane Vandenbusche, *The Gunnison County* (Gunnison, Colo.: B & B Printers, 1980), chs. 13, 14, 18.

10. *Georgetown Courier*, July 8, 1911, March 4 and September 2, 1916, August 4, 1917, January 10, 1919, September 4, 1920; *Silver Standard*, September 13, 1902, December 5, 1903.

11. The inspector's reports for all these mines are found in the archives of the Division of Mines, Department of Natural Resources, Denver.

12. *The Mountain States Mineral Age* (Denver: Mineral Age Publishing, 1922), 27.

13. Robert Service, "The Prospector," in *Best of Robert Service* (New York: Dodd, Mead, 1953), 56.

CHAPTER 11: MUCKING THROUGH DEPRESSION, WAR, AND NEW IDEAS

1. *Colorado Year Book* (Denver: State of Colorado, 1939), 7.

2. David Lavender, *One Man's West* (Lincoln: University of Nebraska Press, 2007 new ed.), 29, 30–31, 33.

3. Student interview of Minnie Lewis for Professor Ken Periman's class, Fort Lewis College.

4. Charles Merrill, Charles Henderson, and O. E. Kiessling, *Small Scale Placer Mines as a Source of Gold Employment* (Philadelphia: U. S. Works Projects Administration, 1937), 28; Charles W. Miller, Jr., *The Automobile Gold Rushes* (Moscow: University of Idaho Press, 1998), 59–60, 86–87, 133.

5. Marshall Sprague, *Money Mountain* (Boston: Little, Brown, 1953), 292.

6. John Joyce, *Annual Report* (Denver: Colorado Bureau of Mines, 1929), 9.

7. Colorado Bureau of Mines, *Annual Report for 1933* (Denver: State of Colorado, 1933), 10, 22; *Durango Weekly Herald*, April 8, 1937; *Durango News*, February 28, 1936.

8. Bill Jones, "Charles A. Chase and the Shenandoah-Dives Mines," in *Mining the Hard Rock in the Silverton San Juans* (Silverton, Colo.: Simpler Way Book Co., 1996), 24–26.

9. *Minerals Yearbook 1942* (Washington, D.C.: Government Printing Office, 1942), 8, 26, 34, 80–82, 317, 322.

10. *Minerals Yearbook 1951* (Washington, D.C.: Government Printing Office, 1952), 145–146.

11. Thomas Hornsby Ferril, *New and Selected Poems* (Westport, Conn.: Greenwood Press, 1952), 109.

12. *Colorado Annual Report 1962* (Denver: Colorado Bureau of Mines, 1962), 62.

13. J. P. Wood, *Report on Mine Tailing Pollution of Clear Creek, Clear Creek-Gilpin Counties, Colorado* (Denver: n.p., 1935), 9, 11, 16, 23, 29–31, 34; *Humphreys Tunnel Mining Co. v. Frank*, 46 Colo. 524, 530–32, 105 P. 1093 (1909); *Wilmore v. Chain O'Mines, Inc.*, 96 Colo. 319, 321, 324, 330–331, 44 P.2d 1024 (1935); *Elk Mountain Pilot*, March 21, 1935. See also *Slide Mines, Inc. v. Left Hand Ditch Co.*, 102 Colo. 69, 77 P.2d 125 (1928).

14. Chase to Lyon, January 30, 1946, January 25, 1950, January 30, 1946, in author's possession; *Silverton Standard*, March 6, 1980.

15. *Mineral and Water Resources of Colorado* (Washington, D.C.: Government Printing Office, 1964), 84.

CHAPTER 12: MINING ON THE DOCKET OF PUBLIC OPINION: THE ENVIRONMENTAL AGE

1. Helen Hunt Jackson, "O-Be-Joyful Creek and Poverty Gulch," *Atlantic Monthly*, December, 1883, 755–756.

2. James Grafton Rogers, *My Rocky Mountain Valley* (Boulder, Colo.: Pruett Press, 1968), 47–48.

3. "Report," *Western Resources Conference* (Golden: Colorado School of Mines, 1971), iii.

4. *Grand Junction Sentinel*, quoted in the *Mining Record*. September 4, 1974; *Mining Record*, January 27, 1982, and February 15, 1984.

5. *A Summary of Mineral Industry Activities in Colorado 1976: Part II Metal-Nonmetal* (Denver: Colorado Division of Mines, 1977), 9, 10–26, 53–57.

6. For the Summitville mess, see *Denver Post* reports from 1889 to 2000. Also see "The Summitville Mine and Its Downstream Effects," U.S. Geological Survey, On-Line Update, July 11, 1995, 1–4; "Report of Review: Region 8 … Ensure Efficient Summitville Superfund Site Cleanup" (Kansas City: EPA, 1996), 5–13; Jennifer Wells, "Canada's Next Billionaire," *Maclean's*, June 3, 1996, 43–44; *High Country News*, January 19, 1998, 6, 10; "Environmental Considerations of Active and Abandoned Mine Lands: Lessons from Summitville, Colorado" (Washington, D.C.: Government Printing Office, 1995), 2–6, 35.

7. *Denver Post*, October 9, 2007.

Remember, the past tells us of ourselves
and gives hints of the future.

Bibliographical Essay

Colorado mining has attracted an increasing number of researchers, historians, writers, and ordinary but interested folk over the past 150 years. It is a fascinating story, one well worth digging into with vigor. While it is impossible to list all the books, articles, and primary source materials that are available, the following are places to start prospecting. The field opening to you is limited only by your own interests.

The chapter footnotes provide a sweeping panorama of sources, each one of which will lead to others. Although it takes time to pursue research through them, Colorado newspapers hold a high-grade deposit for studying mining history on every imaginable subject. Beware of becoming so interested that you deviate from your original search! Olive M. Jones's *Bibliography of Colorado Geology and Mining* (Denver, Colo.: Smith-Brooks, 1914) is a bonanza of articles and books to 1912. Arthur E. Smith, Jr., brings the list nearly up to date in his *Bibliography of Colorado Mining History* (Houston, Tex.: Mineral Design Co., 1993). Clark C. Spence's chapter, "Western Mining," in

Michael P. Malone (ed.), *Historians and the American West* (Lincoln: University of Nebraska Press, 1983), offers a wide selection of articles and books on mining in general and Colorado mining specifically.

Charles W. Henderson's *Mining in Colorado: A History of Discovery, Development and Production* (Washington, D.C.: Government Printing Office, 1926), is unsurpassed for a county-by-county breakdown. United States government publications are a gold mine waiting to be explored, starting in the 1860s and continuing into the twenty-first century. For those unclear on or confused about mining terms, *A Dictionary of Mining, Mineral, and Related Terms* (Washington, D.C.: U.S. Department of the Interior, 1968) is the source to check.

To place Colorado in the overall western mining picture, see Rodman W. Paul, *Mining Frontiers of the Far West* (rev. ed., Albuquerque: University of New Mexico Press, 2001) and William Greever, *The Bonanza West: The Story of the Western Mining Rushes, 1848–1900* (Norman: University of Oklahoma Press, 1963). Otis Young, Jr., has written several books covering a variety of mining methods and topics. Ronald Brown, Richard Lingenfelter, and Mark Wyman have all written books on western miners and mining unions.

All of the following take the reader into these fascinating topics: Clark C. Spence, *Mining Engineers & The American West* (New Haven: Yale University Press, 1970) and *British Investments and the American Mining Frontier, 1860–1901* (Ithaca, N.Y.: Cornell University Press, 1958); Joseph E. King, *A Mine to Make a Mine: Financing the Colorado Mining Industry, 1859–1902* (College Station: Texas A&M University Press, 1977); James E. Fell, Jr., *Ores to Metals: The Rocky Mountain Smelting Industry* (Lincoln: University of Nebraska Press, 1979); and Robert A. Trennert, *Riding the High Wire: Aerial Mine Tramways in the West* (Boulder: University Press of Colorado, 2001).

Muriel Sibell Wolle visited, sketched, interviewed, and wrote; her output of books and articles presents a kaleidoscopic portrait of mining communities, people, districts, and mines. See also the author's collection of books and articles, including *Rocky Mountain Mining Camps* (Bloomington: Indiana University Press, 1967; paperback ed., Lincoln: University Press of Colorado, 1992) and *Colorado Mining: A Photographic History* (Albuquerque: University of New Mexico Press, 1977).

It is amazing how many Colorado mining camps and towns have found a Homer to recount their history. There are far too many to list here without omitting some noteworthy ones. Their readability, reliability, and significance range about as widely as the importance of the communities themselves in Colorado mining history.

Only a few individuals prominent in Colorado mining history have received scholarly attention; some of what has been written is more spectacular than scholarly. For those interested in research in original material, the Denver Public Library, University of Colorado at Boulder Library, Colorado Historical Society, Fort Lewis College Southwest Center, Colorado State Archives, and the Colorado School of Mines Library have individual, company, and a variety of other records pertaining to Colorado mining. There are also smaller libraries and museums throughout the state that contain some mining gems.

In the mining West, the old saying "it takes a mine to run a mine" was often true. In studying Colorado mining history, it is similarly true that it takes a book to introduce the researcher to other books, articles, and primary sources. For that reason, this bibliography shows the way to a handful of resources to tempt one forward, but should never be considered all-inclusive. So go forth and, as another bit of mining lore states, "Tap 'er light." Good luck, and enjoy your journey through this fascinating subject.

Index